LEARNING GUIDE WITH INTEGRATED REVIEW WORKSHEETS

BONNIE ROSENBLATT
Reading Area Community College

THINKING MATHEMATICALLY
SIXTH EDITION

Robert Blitzer
Miami Dade College

PEARSON

Boston   Columbus   Indianapolis   New York   San Francisco   Upper Saddle River
Amsterdam   Cape Town   Dubai   London   Madrid   Milan   Munich   Paris   Montreal   Toronto
Delhi   Mexico City   São Paulo   Sydney   Hong Kong   Seoul   Singapore   Taipei   Tokyo

Reproduced by Pearson from electronic files supplied by the author.

Copyright © 2015, 2011, 2008 Pearson Education, Inc.
Publishing as Pearson, 75 Arlington Street, Boston, MA 02116.

ISBN-13: 978-0-321-91538-2
ISBN-10: 0-321-91538-0

6   17

www.pearsonhighered.com

PEARSON

Table of Contents

Chapter 1
Problem Solving and Critical Thinking

Section 1.1 Inductive and Deductive Reasoning

- **Objective 1 - Understand and use inductive reasoning.**

Solved Problem:

Identify a pattern in each list of numbers. Then use this pattern to find the next number.

a. 3, 9, 15, 21, 27, _____
b. 2, 10, 50, 250, _____
c. 3, 6, 18, 72, 144, 432, _____
d. 1, 9, 17, 3, 11, 19, 5, 13, 21, _____

a. Add 6 each time.
 $3 + 6 = 9$
 $9 + 6 = 15$
 $15 + 6 = 21$
 $21 + 6 = 27$
 $27 + 6 = 33$
 3, 9, 15, 21, 27, <u>33</u>

b. Multiply by 5 each time.
 $2 \times 5 = 10$
 $10 \times 5 = 50$
 $50 \times 5 = 250$
 $250 \times 5 = 1250$
 2, 10, 50, 250, <u>1250</u>

c. Cycle multiplying by 2, 3, 4.
 $3 \times 2 = 6$
 $6 \times 3 = 18$
 $18 \times 4 = 72$
 $72 \times 2 = 144$
 $144 \times 3 = 432$
 $432 \times 4 = 1728$
 $1728 \times 2 = 3456$
 6, 18, 72, 144, 432, 1728, <u>3456</u>

d. Cycle adding 8, adding 8, subtracting 14.
 $1 + 8 = 9$
 $9 + 8 = 17$
 $17 - 14 = 3$
 $3 + 8 = 11$
 $11 + 8 = 19$
 $19 - 14 = 5$
 $5 + 8 = 13$
 $13 + 8 = 21$
 $21 - 14 = 7$
 9, 17, 3, 11, 19, 5, 13, 21, <u>7</u>

Your Turn (exercises 9, 15, 25, pg. 11):

Identify a pattern in each list of numbers. Then use this pattern to find the next number.

8, 12, 16, 20, 24, ___
1, 2, 4, 8, 16, ___
3, 7, 12, 18, 25, 33, ___

2

- **Objective 2 - Understand and use deductive reasoning.**

Solved Problem:

Select a number. Multiply the number by 4. Add 6 to the product. Divide this sum by 2. Subtract 3 from the quotient.

a. Repeat this process for at least four different numbers. Write a conjecture that relates the results of this process to the original number selected.

b. Use the variable n to represent the original number and use deductive reasoning to prove the conjecture in part (a).

a.

Conjecture based on results: The original number is doubled.

Select a number.	4	10	0	3
Multiply the number by 4.	$4 \times 4 = 16$	$10 \times 4 = 40$	$0 \times 4 = 0$	$3 \times 4 = 12$
Add 6 to the product.	$16 + 6 = 22$	$40 + 6 = 46$	$0 + 6 = 6$	$12 + 6 = 18$
Divide this sum by 2.	$22 \div 2 = 11$	$46 \div 2 = 23$	$6 \div 2 = 3$	$18 \div 2 = 9$
Subtract 3 from the quotient.	$11 - 3 = 8$	$23 - 3 = 20$	$3 - 3 = 0$	$9 - 3 = 6$
Summary of results:	$4 \rightarrow 8$	$10 \rightarrow 20$	$0 \rightarrow 0$	$3 \rightarrow 6$

b. Select a number: n

Multiply the number by 4: $4n$

Add 6 to the product: $4n + 6$

Divide this sum by 2: $\dfrac{4n+6}{2} = \dfrac{4n}{2} + \dfrac{6}{2} = 2n+3$

Subtract 3 from the quotient: $2n + 3 - 3 = 2n$

Your Turn (exercise #43, pg. 11):

Describe procedures that are to be applied to numbers. In each exercise,
a. Repeat the procedure for four numbers of your choice. Write a conjecture that relates the result of the process to the original number selected.
b. Use the variable n to represent the original number and use deductive reasoning to prove the conjecture in part (a).

Select a number. Multiply the number by 4. Add 8 to the product. Divide this sum by 2. Subtract 4 from the quotient.

Section 1.2 Estimation, Graphs and Mathematical Models

- **Objective 1 - Use estimation techniques to arrive at an approximate answer to a problem.**

Solved Problem:

You and a friend ate lunch at Ye Olde Cafe. The check for the meal showed soup for $3.40, tomato juice for $2.25, a roast beef sandwich for $5.60, a chicken salad sandwich for $5.40, two coffees totaling $3.40, apple pie for $2.85, and chocolate cake for $3.95.

a. Round the cost of each item to the nearest dollar and obtain an estimate for the food bill.

b. The total bill before tax was given as $29.85. Is this amount reasonable?

a. $\$3.40 + \$2.25 + \$5.60 + \$5.40 + \$3.40$
$$+ \$2.85 + 3.95$$
$\approx \$3 + \$2 + \$6 + \$5 + \$3 + \$3 + 4$
$\approx \$26$

b. The given bill is not reasonable. It is too high.

Your Turn (exercise #23, pg. 26):

Determine the estimate without using a calculator. Then use a calculator to perform the computation necessary to obtain an exact answer. How reasonable is your estimate when compared to the actual answer?

Estimate the total cost of six grocery items if their prices are $3.47, $5.89, $19.98, $2.03, $11.85, and $0.23.

4

- **Objective 2 - Apply estimation techniques to information given by graphs.**

Solved Problem:

Use the line graph in **Figure 1.7** (p. 21) to solve this exercise.

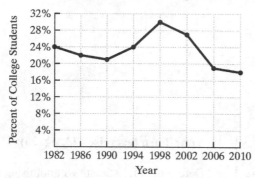

Cigarette Use by U.S. College Students

FIGURE 1.7
Source: Rebecca Donatelle, *Health The Basics*, 10th Edition, Pearson, 2013.

a. Find an estimate for the percentage of college students who smoked cigarettes in 1986.

b. In which four-year period did the percentage of college students who smoked cigarettes increase at the greatest rate?

c. In which years labeled on the horizontal axis did 24% of college students smoke cigarettes?

a. about 22%

b. The greatest rate of increase in the percentage of college students who smoked cigarettes can be found by identifying the portion of the graph with the largest upward slope. This occurs between 1994 and 1998.

c. Approximately 24% of college students smoked cigarettes in 1982 and 1994.

Your Turn (exercise #49, pg. 27-28):

Use the graph on page 27 to solve this exercise.

a. Find an estimate for the percentage of Americans who considered Iraq their country's greatest enemy in 2001.

b. Between which two years did the percentage of Americans who considered Iraq their country's greatest enemy decrease at the greatest rate?

c. In which year did 32% of Americans consider Iraq their country's greatest enemy?

- **Objective 3 - Develop mathematical models that estimate relationships between variables.**

Solved Problem:

The bar graph in **Figure 1.9** (p. 25) shows the average cost of tuition and fees for private four-year colleges, adjusted for inflation.

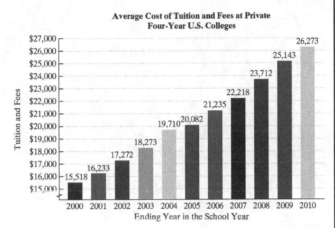

Average Cost of Tuition and Fees at Private Four-Year U.S. Colleges

FIGURE 1.9
Source: The College Board

a. Estimate the yearly increase in tuition and fees. Round to the nearest dollar.

b. Write a mathematical model that estimates the average cost of tuition and fees, T, at private four-year colleges for the school year ending x years after 2000.

c. Use the mathematical model from part (b) to project the average cost of tuition and fees at private four-year colleges for the school year ending in 2014.

a. The yearly increase in tuition and fees can be approximated by dividing the change in tuition and fees by the change in time from 2000 to 2010.

$$\frac{\$26,273 - \$15,518}{2010 - 2000} = \frac{\$10,755}{10} \approx \$1076$$

b. $T = \overbrace{15,518}^{\substack{\text{Cost in} \\ \text{2000}}} + \overbrace{1076}^{\substack{\text{yearly} \\ \text{increase}}} x$

c. 2014 is 14 years after 2000. Thus,

Your Turn (exercise #51, pg. 28):

Use the graph on page 28 to solve the following problem.

a. Estimate the yearly increase in the average atmospheric concentration of carbon dioxide. Express the answer in parts per million, rounded to the nearest hundredth.

b. Write a mathematical model that estimates the average atmospheric concentration of carbon dioxide, C, in parts per million, x years after 1950.

c. If the trend shown by the data continues, use your mathematical model from part (b) to project the average atmospheric concentration of carbon dioxide in 2050.

6

$$T = 15,518 + 1076x$$
$$= 15,518 + 1076(14)$$
$$= \$30,582$$

Section 1.3 Problem Solving

- **Objective 1 – Solve problems using the organization of the four-step problem-solving process.**

Solved Problem:

By paying $350 cash up front and the balance of $45 per month, how long will it take to pay for a computer costing 980?

Step 1: Understand the problem.
We are given the cost of the computer, the amount of cash paid up front, and the amount paid each month. We must determine the number of months it will take to finish paying for the computer.

Step 2: Devise a plan.
Subtract the amount paid in cash from the cost of the computer. This results in the amount still to be paid. Because the monthly payments are $45, divide the amount still to be paid by 45. This will give the number of months required to pay for the computer.

Step 3: Carry out the plan and solve the problem.
The balance is $980 - $350 = $630.$ Now divide the $630 balance by $45, the monthly payment.

$$\$630 \div \frac{\$45}{\text{month}} = \$630 \times \frac{\text{month}}{\$45} = \frac{630 \text{ months}}{45} = 14 \text{ months.}$$

Step 4: Look back and check the answer.
This answer satisfies the conditions of the problem. 14 monthly payments at $45 each gives $14 \times \$45 = \$630.$ Adding in the up front cash payment of $350 gives us $\$630 + \$350 = \$980.$ $980 is the cost of the computer.

Your Turn (exercise #13, pg. 38):

A television sells for $750. Instead of paying the total amount at the time of the purchase, the same television can be bought by paying $100 down and $50 a month for 14 months. How much is saved by paying the total amount at the time of the purchase?

8

Chapter 2
Set Theory

Section 2.1 Basic Set Concepts

- **Objective 1 - Use three methods to represent sets.**

Solved Problem:

Express the set

$O = \{x \mid x \text{ is a positive odd number less than } 10\}$

$O = \{1, 3, 5, 7, 9\}$

Your Turn (exercise #17, pg. 58):

Express this set using the roster method:

$\{x \mid x \text{ is a month that ends with the letters b-e-r}\}$

10

- **Objective 2 - Define and recognize the empty set.**

Solved Problem:

Which one of the following is the empty set?

a. $\{x \mid x$ is a number less than 3 or greater than 5$\}$

b. $\{x \mid x$ is a number less than 3 and greater than 5$\}$

c. Nothing

d. $\{\varnothing\}$

a. not the empty set; Many numbers meet the criteria to belong to this set.

b. the empty set; No numbers meet the criteria, thus this set is empty

c. not the empty set; "nothing" is not a set.

d. not the empty set; This is a set that contains one element, that element is a set.

Your Turn (exercise #33, 39, 41, 45, pg. 58}:

Determine which sets are the empty set.

a. $\{\varnothing, 0\}$

b. $\{x \mid x$ is a U.S. state whose name begins with the letter X$\}$

c. $\{x \mid x < 2$ and $x > 5\}$

d. $\{x \mid x$ is a number less than 2 or greater than 5$\}$

- **Objective 3 - Use the symbols $\in$ and $\notin$.**

Solved Problem:

Determine whether each statement is true or false:

a. $8 \in \{1, 2, 3, ..., 10\}$

b. $r \notin \{a, b, c, z\}$

c. $\{Monday\} \in \{x \mid x$ is a day of the week$\}$

a. true; 8 is an element of the given set.

b. true; r is not an element of the given set.

c. false; {Monday} is a set and the set{Monday} is not an element of the given set.

Your Turn (exercise #49, 53, 63, pg. 58-59}:

Determine whether each statement is true or false:

a. $12 \in \{1, 2, 3, ..., 14\}$

b. $11 \notin \{1, 2, 3, ..., 9\}$

c. $\{3\} \in \{3, 4\}$

12

- **Objective 4 - Apply set notation to sets of natural numbers.**

Solved Problem:

Express each of the following sets using the roster method:

a. Set A is the set of natural numbers less than or equal to 3.

b. Set B is the set of natural numbers greater than 14.

c. $O = \{x \mid x \notin \mathbf{N} \text{ and } x \text{ is odd}\}$

a. $A = \{1, 2, 3\}$

b. $B = \{15, 16, 17, \ldots\}$

c. $O = \{1, 3, 5, \ldots\}$

Your Turn (exercise #19, 21, 25, pg. 58):

Express each of the following sets using the roster method:

a. The set of natural numbers less than 4

b. The set of odd natural numbers less than 13

c. $\{x \mid x \in \mathbf{N} \text{ and } x > 5\}$

- **Objective 5 - Determine a set's cardinal number.**

Solved Problem:

Find the cardinal number of each of the following sets:

a. $A = \{6, 10, 14, 15, 16\}$

b. $B = \{872\}$

c. $C = \{9, 10, 11, \ldots, 15, 16\}$

d. $D = \{\ \}$

a. $n(A) = 5$; the set has 5 elements

b. $n(B) = 1$; the set has only 1 element

c. $n(C) = 8$; Though this set lists only five elements, the three dots indicate 12, 13, and 14 are also elements.

d. $n(D) = 0$ because the set has no elements.

Your Turn (exercise #67, 69, 71, 73, pg. 59}:

Find the cardinal number for each of the following sets:

a. $A = \{17, 19, 21, 23, 25\}$

b. $B = \{2, 4, 6, \ldots, 30\}$

c. $C = \{x \mid x$ is a day of the week that begins with the letter A$\}$

d. $D = \{\text{five}\}$

14

Objective 6 - Recognize equivalent sets.

Solved Problem:

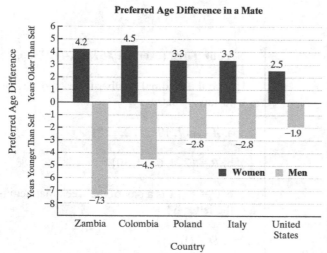

Preferred Age Difference in a Mate

Let
A = the set of the five countries shown in **Figure 2.1** (p. 55).
B = the set of the average number of years men in each of these countries prefer women who are younger than themselves.

Are these sets equivalent? Explain.

No, the sets are not equivalent. Set A has 5 elements yet set B has only 4 elements.

Your Turn (exercise #83a, pg. 59):

Are these sets equivalent?

$A = \{1, 2, 3, 4, 5\}$

$B = \{0, 1, 2, 3, 4\}$

- **Objective 8 - Recognize equal sets.**

Solved Problem:

Determine whether each statement is true or false:

a. {O, L, D} = {D, O, L}

b. {4, 5} = {5, 4, ∅ }

a. true; {O, L, D} = {D, O, L} because the sets contain exactly the same elements.

b. false; The two sets do not contain exactly the same elements.

Your Turn (exercise #83b, 85b, pg. 59):

a. Are these sets equal?

$A = \{1, 2, 3, 4, 5\}$

$B = \{0, 1, 2, 3, 4\}$

b. Are these sets equal?

$A = \{1, 1, 1, 2, 2, 3, 4\}$

$B = \{4, 3, 2, 1\}$

Section 2.2 Subsets

- **Objective 1 - Recognize subsets and use the notation ⊆.**

Solved Problem:

Write ⊆ or ⊄ in each blank to form a true statement:

a. $A = \{1, 3, 5, 6, 9, 11\}$
   $B = \{1, 3, 5, 7\}$

A_____B

b. $A = \{\, x \mid x$ is a letter in the word *roof* $\}$
   $B = \{\, y \mid y$ is a letter in the word *proof* $\}$

A_____B

c. $A = \{\, x \mid x$ is a day of the week$\}$
   $B = \{$ Monday, Tuesday, Wednesday, Thursday, Friday, Saturday, Sunday$\}$

A_____B

a. ⊄ ; because 6, 9, and 11 are not in set B.

b. ⊆ ; because all elements in set A are also in set B.

c. ⊆ ; because all elements in set A are also in set B.

Your Turn (exercise #3, 5, 9, pg. 68}:

Write ⊆ or ⊄ in each blank to form a true statement:

a. $A = \{-3, 0, 3\}$
   $B = \{-3, -1, 1, 3\}$

A_____B

b. $A = \{$Monday, Friday$\}$
   $B = \{$Saturday, Sunday, Monday, Tuesday, Wednesday$\}$

A_____B

c. $A = \{$c, o, n, v, e, r, s, a, t, i, o, n$\}$
   $B = \{$v, o, i, c, e, s, r, a, n, t, o, n$\}$

A_____B

- **Objective 2 - Recognize proper subsets and use the notation ⊂.**

Solved Problem:

Write ⊆, ⊂, or both in each blank to form a true statement:

a. $A = \{2, 4, 6, 8\}$
   $B = \{2, 8, 4, 6, 10\}$

A_____B

b. $A = \{\, x \mid x$ is a person and x lives in Atlanta$\}$
   $B = \{\, x \mid x$ is a person and x lives in Georgia$\}$

A_____B

a. Both ⊆ and ⊂ are correct.

b. Both ⊆ and ⊂ are correct.

Your Turn (exercise #21, 25, pg. 68}:

Determine whether ⊂, ⊆, both, or neither can be placed in each blank to form a true statement:

a. $A = \{0, 2, 4, 6, 8\}$
   $B = \{8, 0, 6, 2, 4\}$

A_____B

b. $A = \{x \mid x$ is a man$\}$
   $B = \{x \mid x$ is a person$\}$

A_____B

18

- **Objective 3 - Determine the number of subsets of a set.**

Solved Problem:

Find the number of distinct subsets and the number of distinct proper subsets for each set:

a. {a, b, c, d}

b. $\{ x \mid x \in \mathbf{N} \text{ and } 3 \leq x \leq 8 \}$

a. 16 subsets, 15 proper subsets
There are 4 elements, which means there are 2^4 or 16 subsets. There are $2^4 - 1$ proper subsets or 15.

b. 64 subsets, 63 proper subsets
There are 6 elements, which means there are 2^6 or 64 subsets. There are $2^6 - 1$ proper subsets or 63.

Your Turn (exercise #63, 67, pg. 68}:

Find the number of distinct subsets and the number of distinct proper subsets for each set:

a. {2, 4, 6, 8, 10, 12}

b. $\{x \mid x \in \mathbf{N} \text{ and } 2 < x < 6\}$

Section 2.3 Venn Diagrams and Set Operations

- **Objective 3 - Use Venn diagrams to visualize relationships between two sets.**

Solved Problem:	**Your Turn (exercise #67, pg. 81}:**
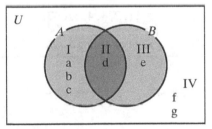	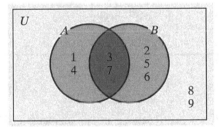
FIGURE 2.11	
Use the Venn diagram in **Figure 2.11** (p. 73) to determine each of the following sets:	Use the Venn diagram above to represent the set *A* in roster form.
a. *A*	
a. {a, b, c, d}	

20

- **Objective 4 - Find the complement of a set.**

Solved Problem:

Let $U = \{a, b, c, d, e\}$ and
$A = \{a, d\}$.

Find A'

$A' = \{b, c, e\}$; those are the elements in U but not in A.

Your Turn (exercise #21, pg. 81}:

Let $U = \{1, 2, 3, 4, 5, 6, 7\}$
$A = \{1, 3, 5, 7\}$
$B = \{1, 2, 3\}$
$C = \{2, 3, 4, 5, 6\}$

Find A'

- **Objective 5 - Find the intersection of two sets.**

Solved Problem:	**Your Turn (exercise #17, 35, 41, pg. 81)**

Find each of the following intersections:

a. $\{1, 3, 5, 7, 10\} \cap \{6, 7, 10, 11\}$

b. $\{1, 2, 3\} \cap \{4, 5, 6, 7\}$

c. $\{1, 2, 3\} \cap \varnothing$

a. $\{1, 3, 5, \underline{7}, \underline{10}\} \cap \{6, \underline{7}, \underline{10}, 11\} = \{7, 10\}$

b. $\{1, 2, 3\} \cap \{4, 5, 6, 7\} = \varnothing$

c. $\{1, 2, 3\} \cap \varnothing = \varnothing$

Let $U = \{1, 2, 3, 4, 5, 6, 7\}$
 $A = \{1, 3, 5, 7\}$
 $B = \{1, 2, 3\}$
 $C = \{2, 3, 4, 5, 6\}$

Find each of the following intersections:

a. $A \cap B$

b. $A \cap \varnothing$

Let $U = \{a, b, c, d, e, f, g, h\}$
 $A = \{a, g, h\}$
 $B = \{b, g, h\}$
 $C = \{b, c, d, e, f\}$

Find $A \cap B$

22

- **Objective 6 - Find the union of two sets.**

Solved Problem:

Find each of the following unions:

a. $\{1, 3, 5, 7, 10\} \cup \{6, 7, 10, 11\}$

b. $\{1, 2, 3\} \cup \{4, 5, 6, 7\}$

c. $\{1, 2, 3\} \cup \varnothing$

a. $\{1, 3, 5, 7, 10\} \cup \{6, 7, 10, 11\}$
   $= \{1, 3, 5, 6, 7, 10, 11\}$

b. $\{1, 2, 3\} \cup \{4, 5, 6, 7\} = \{1, 2, 3, 4, 5, 6, 7\}$

c. $\{1, 2, 3\} \cup \varnothing = \{1, 2, 3\}$

Your Turn (exercise #19, 43, 57, pg. 81):

Let $U = \{1, 2, 3, 4, 5, 6, 7\}$
$A = \{1, 3, 5, 7\}$
$B = \{1, 2, 3\}$
$C = \{2, 3, 4, 5, 6\}$

Find $A \cup B$

Let $U = \{a, b, c, d, e, f, g, h\}$
$A = \{a, g, h\}$
$B = \{b, g, h\}$
$C = \{b, c, d, e, f\}$

a. Find $A \cup B$

b. Find $A \cup \varnothing$

- **Objective 7 - Perform operations with sets.**

Solved Problem:

Given $U = \{a, b, c, d, e\}$,
$\quad A = \{b, c\}$, and
$\quad B = \{b, c, e\}$,

find each of the following sets:

a. $(A \cup B)'$

b. $A' \cap B'$

a. $A \cup B = \{b, c, e\}$
$\quad (A \cup B)' = \{a, d\}$

b. $A' = \{a, d, e\}$
$\quad B' = \{a, d\}$
$\quad A' \cap B' = \{a, d\}$

Your Turn (exercise #27, 29, pg. 81):

Let $\quad U = \{1, 2, 3, 4, 5, 6, 7\}$
$\quad A = \{1, 3, 5, 7\}$
$\quad B = \{1, 2, 3\}$
$\quad C = \{2, 3, 4, 5, 6\}$

Find each of the following sets:

a. $(A \cap C)'$

b. $A' \cup C'$

24

- **Objective 8 - Determine sets involving set operations from a Venn diagram.**

Solved Problem:

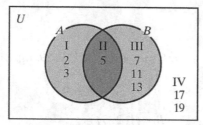

FIGURE 2.17

Use the Venn diagram in **Figure 2.17** (p. 78) to determine each of the following sets.

a. $A \cap B$

b. $(A \cap B)'$

c. $A \cup B$

d. $(A \cup B)'$

e. $A' \cup B$

f. $A \cap B'$

a. {5}; region II

b. {2, 3, 7, 11, 13, 17, 19}; the complement of region II

c. {2, 3, 5, 7, 11, 13}; regions I, II, and III

d. {17, 19}; the complement of regions I, II, and III

e. {5, 7, 11, 13, 17, 19}; the complement of A united with B

f. {2, 3}; A intersected with the complement of B

Your Turn (exercise #71, 75, 77, 81, 87, 89, pg. 81):

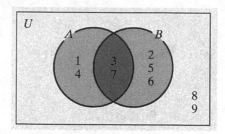

Use the Venn diagram above to represent each set in roster form:

a. $A \cap B$

b. $(A \cup B)'$

c. $A \cap B'$

Use the Venn diagram above to determine each set or cardinality:

a. $A \cup B$

b. $(A \cap B)'$

c. $A' \cap B$

- **Objective 10 - Use the formula for** $n(A \cup B)$.

Solved Problem:

According to factmonster.com, among the U.S. presidents in the White House, 26 had dogs, 11 had cats, and 9 had both dogs and cats. How many U.S. presidents had dogs or cats in the White House?

$$n(A \cup B) = n(A) + n(B) - n(A \cap B)$$
$$= 26 + 11 - 9$$
$$= 28$$

Your Turn (exercise #139, pg. 83):

A winter resort took a poll of its 350 visitors to see which winter activities people enjoyed. The results were as follows: 178 people liked to ski, 154 people liked to snowboard, and 49 people liked to ski and snowboard. How many people in the poll liked to ski or snowboard?

26

Section 2.4 Set Operations and Venn Diagrams with Three Sets

- **Objective 1 - Perform set operations with three sets.**

Solved Problem:

Given

$U = \{a, b, c, d, e, f\}$,
$A = \{a, b, c, d\}$,
$B = \{a, b, d, f\}$,
$C = \{b, c, f\}$,

find each of the following sets:

a. $A \cup (B \cap C)$

b. $(A \cup B) \cap (A \cup C)$

c. $A \cap (B \cup C')$

a. $A \cup (B \cap C) = \{a, b, c, d\} \cup \{b, f\}$
$= \{a, b, c, d, f\}$

b. $(A \cup B) \cap (A \cup C) = \{a, b, c, d, f\} \cap \{a, b, c, d, f\}$
$= \{a, b, c, d, f\}$

c. $A \cap (B \cup C') = \{a, b, c, d\} \cap (\{a, b, d, f\} \cup \{a, d, e\})$
$= \{a, b, c, d\} \cap \{a, b, d, e, f\}$
$= \{a, b, d\}$

Your Turn (exercise #1, 3, 5):

Let $U = \{1, 2, 3, 4, 5, 6, 7\}$
$A = \{1, 3, 5, 7\}$
$B = \{1, 2, 3\}$
$C = \{2, 3, 4, 5, 6\}$

Find each of the following sets:

a. $A \cup (B \cap C)$

b. $(A \cup B) \cap (A \cup C)$

c. $A' \cap (B \cup C')$

- **Objective 2 - Use Venn diagrams with three sets.**

Solved Problem:

Construct a Venn diagram illustrating the following sets:

$A = \{1, 3, 6, 10\}$
$B = \{4, 7, 9, 10\}$
$C = \{3, 4, 5, 8, 9, 10\}$
$U = \{1, 2, 3, 4, 5, 6, 7, 8, 9, 10\}$

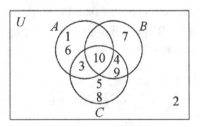

Your Turn (exercise #45, pg. 92):

Construct a Venn diagram illustrating the following sets:

A = {4, 5, 6, 8}
B = {1, 2, 4, 5, 6, 7}
C = {3, 4, 7}
U = {1, 2, 3, 4, 5, 6, 7, 8, 9}

28

• **Objective 3 - Use Venn diagrams to prove equality of sets.**

Solved Problem:

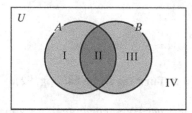

FIGURE 2.22

Use the Venn diagram in **Figure 2.22** (p. 89) to solve this exercise.

a. Which region represents $(A \cup B)'$?

b. Which region represents $A' \cap B'$?

c. Based on parts (a) and (b), what can you conclude?

a. $A \cup B$ is represented by regions I, II, and III. Therefore $(A \cup B)'$ is represented by region IV.

b. A' is represented by regions III and IV.
   B' is represented by regions I and IV.
   Therefore $A' \cap B'$ is represented by region IV.

c. $(A \cup B)' = A' \cap B'$ because they both represent region IV.

Your Turn (exercise #51, pg. 92):

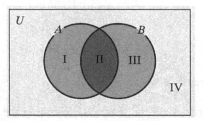

Use the Venn diagram above to solve this exercise.

a. Which region(s) represent(s) $(A \cap B)'$?

b. Which region(s) represent(s) $A' \cap B'$?

c. Based on parts (a) and (b), are $(A \cap B)'$ and $A' \cap B'$ equal for all sets A and B? Explain your answer.

Section 2.5 Survey Problems

- **Objective 1 - Use Venn diagrams to visualize a survey's results.**

Solved Problem:

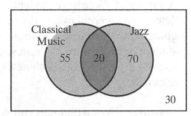

FIGURE 2.27

In a survey on musical tastes, respondents were asked:

Do you listen to classical music? Do you listen to jazz? The survey results are summarized in **Figure 2.27** (p. 97). Use the diagram to answer the following questions.

a. How many respondents listened to classical music?
b. How many respondents listened to jazz?
c. How many respondents listened to both classical music and jazz?
d. How many respondents listened to classical music or jazz?
e. How many respondents listened to classical music but not jazz?
f. How many respondents listened to jazz but not classical music?
g. How many respondents listened to neither classical music nor jazz?
h. How many people were surveyed?

a. $55 + 20 = 75$
b. $20 + 70 = 90$
c. 20
d. $55 + 20 + 70 = 145$
e. 55
f. 70
g. 30
h. $55 + 20 + 70 + 30 = 175$

Your Turn (exercise #33-38, pg. 104):

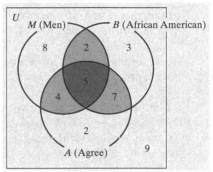

A poll asked respondents if they agreed with the statement "Colleges should reserve a certain number of scholarships exclusively for minorities and women." Hypothetical results of the poll are tabulated in the Venn diagram. Use these cardinalities to solve the below exercises.

a. How many respondents agreed with the statement?
b. How many respondents disagreed with the statement?
c. How many women agreed with the statement?
d. How many people who are not African American agreed with the statement?
e. How many women who are not African American disagreed with the statement?
f. How many men who are not African American disagreed with the statement?

- **Objective 2 - Use survey results to complete Venn diagrams and answer questions about the survey.**

Solved Problem:

In a Gallup poll, 2000 U.S. adults were selected at random and asked to agree or disagree with the following statement:

Job opportunities for women are not equal to those for men.

The results of the survey showed that 1190 people agreed with the statement. 700 women agreed with the statement. *Source: The People's Almanac*

If half the people surveyed were women,

a. How many men agreed with the statement?
b. How many men disagreed with the statement?

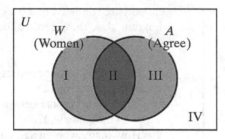

Start by placing 700 in region II.
Next place $1190 - 700$ or 490 in region III.
Since half of those surveyed were women, place $1000 - 700$ or 300 in region I.
Finally, place $2000 - 300 - 700 - 490$ or 510 in region IV.

a. 490 men agreed with the statement and are represented by region III.

b. 510 men disagreed with the statement and are represented by region IV.

Your Turn (exercise #43a, 43c, pg, 105):

A survey of 75 college students was taken to determine where they got the news about what's going on in the world. Of those surveyed, 29 students got the news from newspapers, 43 from television, and 7 from both newspapers and television.

Construct a Venn diagram and determine the cardinality for each region. Use the completed Venn diagram to answer the following questions.

a. Of those surveyed, how many got the news only from newspapers?

b. Of those surveyed, how many got the news from newspapers or television?

Chapter 3
Logic

Section 3.1 Statements, Negations, and Quantified Statements

- Objective 3 - Form the negation of a statement.

Solved Problem:

Form the negation of each statement:

a. Paris is the capital of Spain.

b. July is not a month.

a. Paris is not the capital of Spain.
b. July is a month.

Your Turn (exercise #15, 19, pg. 120):

Form the negation of each statement:

a. It is raining.

b. It is not true that chocolate in moderation is good for the heart.

32

- Objective 4 - Express negations using symbols.

Solved Problem:

Let p and q represent the following statements:

p: Paris is the capital of Spain.
q: July is not a month.

Express each of the following statements symbolically:

a. Paris is not the capital of Spain.
b. July is a month.

a. $\sim p$

b. $\sim q$

Your Turn (exercise #21, 23, pg. 120):

Let p, q, r, and s represent the following statements:

p: One works hard.
q: One succeeds.
r: The temperature outside is not freezing.
s: It is not true that the heater is working.

Express each of the following statements symbolically:

a. One does not work hard.

b. The temperature outside is freezing.

- Objective 5 - Translate a negation represented by symbols into English.

Solved Problem:

Let q represent the following statement:

q: Chicago O'Hare is the world's busiest airport.

Express the symbolic statement $\sim q$ in words.

Chicago O'Hare is not the world's busiest airport.

Your Turn (exercise #25, pg. 120):

Let p represent the following statement:

p: Listening to classical music makes infants smarter.

Express the symbolic statement $\sim p$ in words.

34

- Objective 7 - Write negations of quantified statements.

Solved Problem:

The board of supervisors told us, "All new tax dollars will be used to improve education." I later learned that the board of supervisors never tells the truth.

What can I conclude?
Express the conclusion in two equivalent ways.

Some new tax dollars will not be used to improve education.

At least one new tax dollar will not be used to improve education.

Your Turn (exercise #31b, pg. 120):

Write the negation of "Some students are business majors."

Section 3.2 Compound Statements and Connectives

- Objective 1 - Express compound statements in symbolic form.

Solved Problem:

Use the representations in Example 3 (p. 124 - 125) to write each compound statement below in symbolic form:

Let p and q represent the following simple statements:

p: A person is a father.
q: A person is a male.

a. If a person is not a father, then that person is not a male.

b. If a person is a male, then that person is not a father.

a. $\sim p \rightarrow \sim q$
b. $q \rightarrow \sim p$

Your Turn (exercise #11, 13, pg. 132):

Let p and q represent the following simple statements:

p: This is an alligator.
q: This is a reptile.

Write each compound statement in symbolic form:

a. If this is an alligator, then this is a reptile.

b. If this is not an alligator, then this is not a reptile.

36

- Objective 2 - Express symbolic statements with parentheses in English.

Solved Problem:

Let p, q, and r represent the following simple statements:

p: The plant is fertilized.
q: The plant is not watered.
r: The plant wilts.

Write each of the symbolic statements in words:

a. $(p \land \sim q) \rightarrow \sim r$

b. $p \land (\sim q \rightarrow \sim r)$

a. If the plant is fertilized and watered, then the plant does not wilt.

b. The plant is fertilized, and if the plant is watered then it does not wilt.

Your Turn (exercise #67, 69, pg. 133):

Write each symbolic statement in words. If a symbolic statement is given without parentheses, place them, as needed, before and after the most dominant connective and then translate into English.

a. $(p \land q) \rightarrow r$

b. $p \land (q \rightarrow r)$

- Objective 3 - Use the dominance of connectives.

Solved Problem:

Write each compound statement below in symbolic form:

a. If there is too much homework or a teacher is boring then I do not take that class.

b. There is too much homework, or if a teacher is boring then I do not take that class.

p: There is too much homework.
q: A teacher is boring
r: I take the class.

a. $(p \vee q) \rightarrow \sim r$

b. $p \vee (q \rightarrow \sim r)$

Your Turn (exercise #85, 87, pg. 133):

Write each compound statement in symbolic form:

a. I miss class if and only if it's not true that both I like the teacher and the class is interesting.

b. If I like the teacher, I do not miss class and if and only if the course is interesting.

38

Section 3.3 Truth Tables for Negation, Conjunction, and Disjunction

- Objective 1 - Use the definitions of negation, conjunction, and disjunction.

| **Solved Problem:** | **Your Turn (exercise #3, 5, 9, 11, pg. 147):** |

Solved Problem:

Let p and q represent the following statements:
$p : 3 + 5 = 8$
$q : 2 \times 7 = 20$.

Determine the truth value for each statement:
a. $p \wedge q$ b. $p \wedge \sim q$
c. $\sim p \vee q$ d. $\sim p \vee \sim q$

$p : 3 + 5 = 8$ is true
$q : 2 \times 7 = 20$ is false

a. $p \wedge q$
    T ∧ F
     F
b. $p \wedge \sim q$
    T ∧ ~F
    T ∧ T
     T
c. $\sim p \vee q$
    ~T ∨ F
    F ∨ F
     F
d. $\sim p \vee \sim q$
    ~T ∨ ~F
    F ∨ T
     T

Your Turn (exercise #3, 5, 9, 11, pg. 147):

Let p and q represent the following statements:

$p : 4 + 6 = 10$
$q : 5 \times 8 = 80$

Determine the truth value for each statement:

a. $p \wedge q$

b. $\sim p \wedge q$

c. $q \vee p$

d. $p \vee \sim q$

- Objective 2 - Construct truth tables.

Solved Problem:

Construct a truth table for $\left(p \wedge \sim q \right) \vee \sim p$ to determine when the statement is true and when the statement is false.

$(p \wedge \sim q) \vee \sim p$

p	q	$\sim p$	$\sim q$	$p \wedge \sim q$	$(p \wedge \sim q) \vee \sim p$
T	T	F	F	F	F
T	F	F	T	T	T
F	T	T	F	F	T
F	F	T	T	F	T

Your Turn (exercise #29, pg. 148):

Construct a truth table for $(p \vee q) \wedge \sim p$.

40

• Objective 3 - Determine the truth value of a compound statement for a specific case.

Solved Problem:

Distribution of Looks in the United States

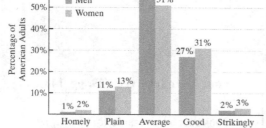

FIGURE 3.2
Source: Time, August 22, 2011

Use the information in the bar graph in **Figure 3.2** (p. 146) to determine the truth value for the following statement:

Two percent of American women are homely or more than half are good looking, and it is not true that 5% of American men are strikingly attractive.

p: Two percent of American women are homely.
q: More than half of American women are good looking.
r: 5% of American men are strikingly attractive.

p is true, *q* is false, *r* is false

$$(p \vee q) \wedge \sim r$$
$$(T \vee F) \wedge \sim F$$
$$T \wedge T$$
$$T$$

The statement is True.

Your Turn (exercise #83, pg. 149):

The bar graph shows the careers named as probable by U.S. college freshmen in 2010.

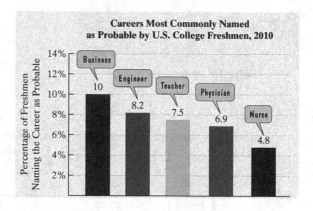

Write the below statement in symbolic form. Then use the information in the graph to determine the truth value of the compound statement.

7.5% named teaching or 6.9% named nursing, and it is not true that 12% named business.

Section 3.4 Truth Tables for the Conditional and the Biconditional

- Objective 2 - Construct truth tables for conditional statements.

Solved Problem:

Construct a truth table for $[(p \rightarrow q) \wedge \sim q] \rightarrow \sim p$ and show that the compound statement is a tautology.

$[(p \rightarrow q) \wedge \sim q] \rightarrow \sim p$ is a tautology because the final column is always true.

p	q	$\sim p$	$\sim q$	$p \rightarrow q$	$(p \rightarrow q) \wedge \sim q$	$[(p \rightarrow q) \wedge \sim q] \rightarrow \sim p$
T	T	F	F	T	F	T
T	F	F	T	F	F	T
F	T	T	F	T	F	T
F	F	T	T	T	T	T

Your Turn (exercise #37, pg. 159):

Use a truth table to determine whether the below statement is a tautology, a self-contradiction, or neither.

$[(p \vee q) \wedge p] \rightarrow \sim q$

42

- Objective 4 - Construct truth tables for biconditional statements.

Solved Problem:

Construct a truth table for
$(p \vee q) \leftrightarrow (\sim p \rightarrow q)$ to determine whether
the statement is a tautology.

$(p \vee q) \leftrightarrow (\sim p \rightarrow q)$ is a tautology because all cases are true.

p	q	$\sim p$	$p \vee q$	$\sim p \rightarrow q$	$(p \vee q) \leftrightarrow (\sim p \rightarrow q)$
T	T	F	T	T	T
T	F	F	T	T	T
F	T	T	T	T	T
F	F	T	F	F	T

Your Turn (exercise #47, pg. 160):

Use a truth table to determine whether the
below statement is a tautology, a self-
contradiction, or neither.

$(p \rightarrow q) \leftrightarrow (\sim p \vee q)$

- Objective 5 - Determine the truth value of a compound statement for a specific case.

Solved Problem:

Consider the underlined claim in the letter in Example 5 (p. 157):

If your Super Million Dollar Prize Entry Number matches the winning preselected number and you return the number before the deadline stated below, you will win $1,000,000.00.

Suppose that your number actually matches the winning preselected number, you do not return the number, and you win nothing. Under these conditions, determine the claim's truth value.

$(p \wedge q) \rightarrow r$
$(T \wedge F) \rightarrow F$
$\quad F \rightarrow F$
$\quad\quad T$

Under these conditions, the claim is true.

Your Turn (exercise #71, pg. 160):

Determine the truth value for the statement below when p is false, q is true, and r is false.

$(\sim p \wedge q) \leftrightarrow \sim r$

44

- Objective 1 - Use a truth table to show that statements are equivalent.

Solved Problem:

Show that $p \vee q$ and $\sim q \to p$ are equivalent.

$p \vee q$ and $\sim q \to p$ are equivalent.

p	q	$\sim q$	$p \vee q$	$\sim q \to p$
T	T	F	T	T
T	F	T	T	T
F	T	F	T	T
F	F	T	F	F

Your Turn (exercise #3, pg. 170):

Use a truth table to show that $\sim p \to q$ and $q \to \sim p$ are equivalent.

- Objective 2 - Write the contrapositive for a conditional statement.

Solved Problem:

Write the contrapositive for each of the following statements:

a. If you can read this, then you're driving too closely.

b. If you do not have clean underwear, it's time to do the laundry.

c. If all students are honest, then supervision during exams is not required.

d. $\sim(p \vee r) \rightarrow \sim q$

a. If you're not driving too closely, then you can't read this.

b. If it's not time to do the laundry, then you have clean underwear.

c. If supervision during exams is required, then some students are not honest.

d. $q \rightarrow (p \vee r)$

Your Turn (exercise #19, 21, 25, 29, pg. 170):

Write the contrapositive for each of the following statements:

a. If I am in Chicago, then I am in Illinois.

b. If the stereo is playing, then I cannot hear you.

c. If the president is telling the truth, then all troops were withdrawn.

d. $\sim q \rightarrow \sim r$

46

- Objective 3 - Write the converse and inverse of a conditional statement.

Solved Problem:

Write the converse and inverse of the following statement:

If you are in Iran, then you don't see a Club Med.

Converse:
If you don't see a Club Med, then you are in Iran.

Inverse:
If you are not in Iran, then you see a Club Med.

Your Turn (exercise #21, pg. 170):

Write the converse and inverse of the following statement:

If the stereo is playing, then I cannot hear you.

Section 3.6 Negations of Conditional Statements and De Morgan's Laws

- Objective 1 - Write the negation of a conditional statement.

Solved Problem:

Write the negation of

If you do not have a fever, you do not have the flu.

You do not have a fever and you have the flu.

Your Turn (exercise #1, pg. 178):

Write the negation of

If I am in Los Angeles, then I am in California.

48

- Objective 2 - Use De Morgan's laws.

Solved Problem:

Write a statement that is equivalent to

It is not true that Bart Simpson and Tony Soprano are cartoon characters.

Bart Simpson is not a cartoon character or Tony Soprano is not a cartoon character.

Your Turn (exercise #11, pg. 178):

Use DeMorgan's laws to write a statement that is equivalent to

It is not true that Australia and China are both islands.

Section 3.7 Arguments and Truth Tables

- Objective 1 - Use truth tables to determine validity.

Solved Problem:

Use a truth table to determine whether the following argument is valid or invalid:

I study for 5 hours or I fail.
I did not study for 5 hours.
Therefore, I failed.

The argument is valid.
p: I study for 5 hours.
q: I fail.

$p \vee q$

$\sim p$

$\therefore q$

p	q	$\sim p$	$p \vee q$	$(p \vee q) \wedge \sim p$	$[(p \vee q) \wedge p] \to q$
T	T	F	T	F	T
T	F	F	T	F	T
F	T	T	T	T	T
F	F	T	F	F	T

Your Turn (exercise #17, pg. 190):

Use a truth table to determine whether the following argument is valid or invalid:

There must be a dam or there is flooding.
This year there is flooding.
Therefore, this year there is no dam.

50

- Objective 2 - Recognize and use forms of valid and invalid arguments.

Solved Problem:

Determine whether each argument is valid or invalid.

a. The emergence of democracy is a cause for hope or environmental problems will overshadow any promise of a bright future. Environmental problems will not overshadow any promise of a bright future. Therefore, the emergence of democracy is a cause for hope.

b. If the defendant's DNA is found at the crime scene, then we can have him stand trial. He is standing trial. Consequently, we found evidence of his DNA at the crime scene.

c. If you mess up, your self-esteem goes down. If your self-esteem goes down, everything else falls apart. So, if you mess up, everything else falls apart.

a. $p \lor q$

$\dfrac{\sim q}{\therefore p}$

This argument is valid by Disjunctive Reasoning.

b. $p \to q$

$\dfrac{q}{\therefore p}$

This argument is invalid by Fallacy of the Converse.

c. $p \to q$

$\dfrac{q \to r}{\therefore p \to r}$

This argument is valid by Transitive Reasoning.

Your Turn (exercise #23, 29, 31, pg. 191):

Determine whether each argument is valid or invalid.

a. We criminalize drugs or we damage the future of young people.
   We will not damage the future of young people.
   Therefore, we criminalize drugs.

b. If I'm tired, I'm edgy.
   If I'm edgy, I'm nasty.
   Therefore, if I'm nasty, I'm tired.

c. If Tim and Janet play, then the team wins.
   Tim played and the team did not win.
   Therefore, Janet did not play.

Section 3.8 Arguments and Euler Diagrams

- Objective 1 - Use Euler diagrams to determine validity.

Solved Problem:

Use Euler diagrams to determine whether the following argument is valid or invalid:

All U.S. voters must register.
All people who register must be U.S. citizens.

Therefore, all U.S. voters are U.S. citizens.

The argument is valid.

The argument is valid.

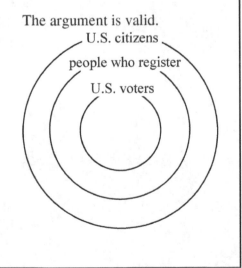

Your Turn (exercise #1, pg. 202):

Fill in the blank so that the resulting statement is true. Refer to parts (a) through (d) in the following figure.

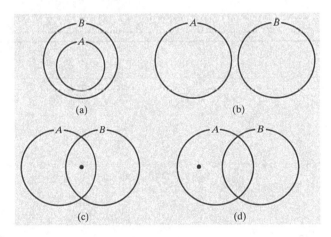

The figure in part (a) illustrates the quantified statement
_____.

52

Chapter 4
Number Representation and Calculation

Section 4.1 Inductive and Deductive Reasoning

- **Objective 2 - Write a Hindu-Arabic numeral in expanded form.**

Solved Problem:

Write each of the following in expanded form:

a. 4026

b. 24,232

a. $4026 =$
$(4 \times 10^3) + (0 \times 10^2) + (2 \times 10^1) + (6 \times 1) =$
$(4 \times 1000) + (0 \times 100) + (2 \times 10) + (6 \times 1)$

b. $24,232 =$
$(2 \times 10^4) + (4 \times 10^3) + (2 \times 10^2) + (3 \times 10^1) + (2 \times 1) =$
$(2 \times 10,000) + (4 \times 1000) + (2 \times 100) + (3 \times 10) + (2 \times 1)$

Your Turn (exercise #13, 19, pg. 218):

Write each Hindu-Arabic numeral in expanded form:

a. 703

b. 34,569

54

- **Objective 3 - Express a number's expanded form as a Hindu-Arabic numeral.**

Solved Problem:

Express each expanded form as a Hindu-Arabic numeral:

a. $\left(6\times10^3\right)+\left(7\times10^1\right)+\left(3\times1\right)$

b. $\left(8\times10^4\right)+\left(9\times10^2\right)$

a. $6000+70+3=6073$

b. $80,000+900=80,900$

Your Turn (exercise #27, 29, pg. 218):

Express each expanded form as a Hindu-Arabic numeral:

a. $\left(5\times10^5\right)+\left(2\times10^4\right)+\left(8\times10^3\right)$
   $+\left(7\times10^2\right)+\left(4\times10^1\right)+\left(3\times1\right)$

b. $\left(7\times10^3\right)+\left(0\times10^2\right)+\left(0\times10^1\right)+\left(2\times1\right)$

- **Objective 4 - Understand and use the Babylonian numeration system.**

Solved Problem:

Write each Babylonian numeral as a Hindu-Arabic numeral:

a. ∨∨∨   <<   <<< ∨

b. ∨∨   <.   <.∨∨   ∨

Your Turn (exercise #39, 45, pg. 218):

Write each Babylonian numeral as a Hindu-Arabic numeral.

a. ∨∨∨   < ∨∨   ∨∨∨

b. ∨∨∨   ∨∨   ∨   ∨

a. First, represent the numeral in each place value as a Hindu-Arabic numeral. Then multiply them by their place values and find the sum.

∨∨∨   <<   <<< ∨
 ↓   ↓   ↓

$(3 \times 60^2) + (20 \times 60) + (31 \times 1)$

$= (3 \times 3600) + (20 \times 60) + (31 \times 1)$

$= 10,800 + 1200 + 31$

$= 12,031$

b. First, represent the numeral in each place value as a Hindu-Arabic numeral. Then multiply them by their place values and find the sum.

∨∨   <   <∨∨   ∨
 ↓   ↓   ↓   ↓

$(2 \times 60^3) + (10 \times 60^2) + (12 \times 60) + (1 \times 1)$

$= (2 \times 216,000) + (10 \times 3600) + (12 \times 60) + (1 \times 1)$

$= 432,000 + 36,000 + 720 + 1$

$= 468,721$

56

- **Objective 5 - Understand and use the Mayan numeration system.**

Solved Problem:

Write each Mayan numeral as a Hindu-Arabic numeral:

a. (Mayan symbols)   b. (Mayan symbols)

Your Turn (exercise #49, 55, pg. 218):

Write each Mayan numeral as a Hindu-Arabic numeral.

a. (Mayan symbols)

b. (Mayan symbols)

a. First, represent the numeral in each place value as a Hindu-Arabic numeral. Then multiply them by their place values and find the sum.

$$11 \times 18 \times 20^2 = 11 \times 7200 = 79,200$$
$$3 \times 18 \times 20^1 = 3 \times 360 = 1080$$
$$0 \times 20 = 0 \times 20 = 0$$
$$13 \times 1 = 13 \times 1 = \underline{\quad 13}$$
$$80,293$$

b. First, represent the numeral in each place value as a Hindu-Arabic numeral. Then multiply them by their place values and find the sum.

$$2 \times 18 \times 20^3 = 2 \times 144,000 = 288,000$$
$$0 \times 18 \times 20^2 = 0 \times 7200 = 0$$
$$6 \times 18 \times 20^1 = 6 \times 360 = 2160$$
$$16 \times 20 = 16 \times 20 = 320$$
$$10 \times 1 = 10 \times 1 = \underline{\quad 10}$$
$$290,490$$

Copyright © 2015 Pearson Education, Inc.

Section 4.2 Number Bases in Positional Systems

- **Objective 1 - Change numerals in bases other than ten to base ten.**

Solved Problem:

Convert 3422_{five} to base ten.

$$3422_{\text{five}} = (3 \times 5^3) + (4 \times 5^2) + (2 \times 5^1) + (2 \times 1)$$
$$= (3 \times 5 \times 5 \times 5) + (4 \times 5 \times 5) + (2 \times 5) + (2 \times 1)$$
$$= 375 + 100 + 10 + 2$$
$$= 487$$

Your Turn (exercise #9, pg. 226):

Convert 2035_{six} to base ten.

58

- **Objective 2 - Change base ten numerals to numerals in other bases.**

Solved Problem:

Convert the base ten numeral 365 to a base seven numeral.

The place values in base 7 are
$...7^4, 7^3, 7^2, 7^1, 1$ or $...2401, 343, 49, 7, 1$

$$\begin{array}{r} 1 \\ 343\overline{)365} \\ \underline{343} \\ 22 \end{array} \qquad \begin{array}{r} 0 \\ 49\overline{)22} \\ \underline{0} \\ 22 \end{array} \qquad \begin{array}{r} 3 \\ 7\overline{)22} \\ \underline{21} \\ 1 \end{array}$$

$$365_{ten} = (1 \times 343) + (0 \times 49) + (3 \times 7) + (1 \times 1)$$
$$= (1 \times 7^3) + (0 \times 7^2) + (3 \times 7^1) + (1 \times 1)$$
$$= 1031_{seven}$$

Your Turn (exercise #35, pg. 226):

Convert the base ten numeral 108 to a base four numeral.

Section 4.3 Computation in Positional Systems

- **Objective 1 - Add in bases other than ten.**

Solved Problem:

Add:

$$\begin{array}{r} 32_{\text{five}} \\ +44_{\text{five}} \\ \hline \end{array}$$

$$\begin{array}{r} \overset{1}{3}2_{\text{five}} \\ +44_{\text{five}} \\ \hline 131_{\text{five}} \end{array}$$

$$2+4 = 6 = (1\times 5^1)+(1\times 1) = 11_{\text{five}}$$
$$1+3+4 = 8 = (1\times 5^1)+(3\times 1) = 13_{\text{five}}$$

Your Turn (exercise #1, pg. 234):

Add:

$$\begin{array}{r} 23_{\text{four}} \\ + 13_{\text{four}} \\ \hline \end{array}$$

60

- **Objective 2 - Subtract in bases other than ten.**

Solved Problem:

Subtract:

$$41_{\text{five}}$$
$$-\ 23_{\text{five}}$$

$$\overset{3\ 6}{4\,1}_{\text{five}}$$
$$\underline{-23_{\text{five}}}$$
$$13_{\text{five}}$$

Your Turn (exercise #13, pg. 234):

Subtract:

$$32_{\text{four}}$$
$$-\ 13_{\text{four}}$$

- **Objective 3 - Multiply in bases other than ten.**

Solved Problem:

Multiply:

$$45_{seven}$$
$$\times\ 3_{seven}$$

$$\overset{2}{4}5_{seven}$$
$$\underline{\times\ 3_{seven}}$$
$$201_{seven}$$

$$3 \times 5 = 15 = (2 \times 7^1) + (1 \times 1) = 21_{seven}$$
$$(3 \times 4) + 2 = 14 = (2 \times 7^1) + (0 \times 1) = 20_{seven}$$

Your Turn (exercise #25, pg. 234):

Multiply:

$$25_{six}$$
$$\underline{\times\ 4_{six}}$$

62

- **Objective 4 - Divide in bases other than ten.**

Solved Problem:

TABLE 4.5		Multiplication: Base Four		
×	**0**	**1**	**2**	**3**
0	0	0	0	0
1	0	1	2	3
2	0	2	10	12
3	0	3	12	21

Use **Table 4.5** (p. 232), showing products in base four, to perform the following division:

$$2_{\text{four}} \overline{)112_{\text{four}}}$$

$$
\begin{array}{r}
23 \\
2_{\text{four}} \overline{)112_{\text{four}}} \\
\underline{10} \\
12 \\
\underline{12} \\
0
\end{array}
$$

$$23_{\text{four}}$$

Your Turn (exercise #37, pg. 234):

MULTIPLICATION: BASE FIVE

×	**0**	**1**	**2**	**3**	**4**
0	0	0	0	0	0
1	0	1	2	3	4
2	0	2	4	11	13
3	0	3	11	14	22
4	0	4	13	22	31

Use the table, showing products in base five, to perform the following division:

$$3_{\text{five}} \overline{)224_{\text{five}}}$$

Section 4.4 Looking Back at early Numeration Systems

- **Objective 1 - Understand and use the Egyptian system.**

Solved Problem:

Hindu-Arabic Numeral	Egyptian Numeral	Description
1			Staff
10	∩	Heel bone
100	ම	Spiral
1000	⚱	Lotus blossom
10,000	⌐	Pointing finger
100,000	⌒	Tadpole
1,000,000	⚇	Astonished person

TABLE 4.6 Egyptian Hieroglyphic Numerals

Using **Table 4.6** (p. 236), write the following numeral as a Hindu- Arabic numeral:

$$100,000 + 100,000 + 100,000 + 100 + 100 + 10 + 10 + 1 + 1 = 300,222$$

Your Turn (exercise #3, pg. 241):

Using **Table 4.6** (p. 236), write the following numeral as a Hindu-Arabic numeral:

64

- **Objective 2 - Understand and use the Roman system.**

Solved Problem:

Using **Table 4.7** (p. 238), write MCDXLVII as a Hindu-Arabic numeral.

$$\text{MCDXLVII} = \overbrace{1000}^{M} + \overbrace{(500-100)}^{CD} + \overbrace{(50-10)}^{XL} + \overset{V}{5} + \overset{I}{1} + \overset{I}{1}$$
$$= 1000 + 400 + 40 + 5 + 1 + 1$$
$$= 1447$$

Your Turn (exercise #25, pg. 241):

Using **Table 4.7** (p. 238), write MDCXXI as a Hindu-Arabic numeral.

TABLE 4.7 Roman Numerals (repeated)							
Roman numeral	I	V	X	L	C	D	M
Hindu-Arabic numeral	1	5	10	50	100	500	1000

- **Objective 3 - Understand and use the traditional Chinese system.**

Solved Problem:

Using **Table 4.8** (p. 239), write 2693 as a Chinese numeral.

$$2693 = 2000 + 600 + 90 + 3$$

二
千
六
百
九
十
三

Your Turn (exercise #45, pg. 242):

Write 583 as a traditional Chinese numeral.

TABLE 4.8 Traditional Chinese Numerals												
Traditional Chinese numerals	一	二	三	四	五	六	七	八	九	十	百	千
Hindu-Arabic numerals	1	2	3	4	5	6	7	8	9	10	100	1000

66

- **Objective 4 - Understand and use the Ionic Greek system.**

Solved Problem:

Using **Table 4.9** (p. 240), write $\omega\pi\varepsilon$ as a Hindu-Arabic numeral.

$$\omega\pi\varepsilon = 800 + 80 + 5 = 885$$

Your Turn (exercise #51, pg. 242):

Using **Table 4.9** (p. 240), write $\sigma\lambda\delta$ as a Hindu-Arabic numeral.

TABLE 4.9		Ionic Greek Numerals						
1	α	alpha	10	ι	iota	100	ρ	rho
2	β	beta	20	κ	kappa	200	σ	sigma
3	γ	gamma	30	λ	lambda	300	τ	tau
4	δ	delta	40	μ	mu	400	υ	upsilon
5	ε	epsilon	50	ν	nu	500	ϕ	phi
6	ι	vau	60	ξ	xi	600	χ	chi
7	ζ	zeta	70	o	omicron	700	ψ	psi
8	η	eta	80	π	pi	800	ω	omega
9	θ	theta	90	Q	koph	900	π	sampi

Chapter 5
Number Theory and the Real Number System

Section 5.1 Number Theory: Prime and Composite Numbers

- **Objective 1 - Determine divisibility.**

Solved Problem:

Which one of the following statements is true?
a. $8 \mid 48{,}324$

b. $6 \mid 48{,}324$

c. $4 \nmid 48{,}324$

The statement given in part (b) is true.

a. false, 8 does not divide 48,324 because 8 does not divide 324.

b. true, 6 divides 48,324 because both 2 and 3 divide 48,324. 2 divides 48,324 because the last digit is 4. 3 divides 48,324 because the sum of the digits, 21, is divisible by 3.

c. false, 4 *does* divide 48,324 because the last two digits form 24 which is divisible by 4.

Your Turn (exercise #11, 15, 19, pg. 256):

Use a calculator to determine whether each statement is true or false. If the statement is true, explain why this is so using one of the rules of divisibility in **Table 5.1** on page 249.

a. $3 \mid 5958$

b. $5 \mid 38{,}814$

c. $8 \mid 20{,}104$

68

- **Objective 2 - Write the prime factorization of a composite number.**

Solved Problem:

Find the prime factorization of 120.

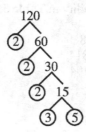

$120 = 2^3 \cdot 3 \cdot 5$

Your Turn (exercise #31, pg. 256):

Find the prime factorization of 500.

- **Objective 3 - Find the greatest common divisor of two numbers.**

Solved Problem:

Find the greatest common divisor of 225 and 825.

$225 = 3^2 \cdot 5^2$
$825 = 3 \cdot 5^2 \cdot 11$

Greatest Common Divisor: $3 \cdot 5^2 = 75$

Your Turn (exercise #53, pg. 256):

Find the greatest common divisor of 342 and 380.

70

- **Objective 4 - Solve problems using the greatest common divisor.**

Solved Problem:

A choral director needs to divide 192 men and 288 women into all-male and all-female singing groups so that each group has the same number of people. What is the largest number of people that can be placed in each singing group?

$192 = 2^6 \cdot 3$

$288 = 2^5 \cdot 3^2$

Greatest Common Divisor: $2^5 \cdot 3 = 96$

The largest number of people that can be placed in each singing group is 96.

Your Turn (exercise #91, pg. 257):

A relief worker needs to divide 300 bottles of water and 144 cans of food into groups that each contain the same number of items. Also, each group must have the same type of item (bottled water or canned food). What is the largest number of relief supplies that can be put in each group?

- **Objective 5 - Find the least common multiple of two numbers.**

Solved Problem:

Find the least common multiple of 18 and 30.

$18 = 2 \cdot 3^2$
$30 = 2 \cdot 3 \cdot 5$
Least common multiple is: $90 = 2 \cdot 3^2 \cdot 5$

Your Turn (exercise #61, pg. 256):

Find the least common multiple of 60 and 108.

72

- **Objective 6 - Solve problems using the least common multiple.**

Solved Problem:

A movie theater runs two documentary films continuously. One documentary runs for 40 minutes and a second documentary runs for 60 minutes. Both movies begin at 3:00 p.m. When will the movies begin again at the same time?

$40 = 2^3 \cdot 5$

$60 = 2^2 \cdot 3 \cdot 5$

Least common multiple is: $120 = 2^3 \cdot 3 \cdot 5$
It will be 120 minutes, or 2 hours, until both movies begin again at the same time.
The time will be 5:00 PM.

Your Turn (exercise #95, pg. 257):

You and your brother both work the 4:00 P.M. to midnight shift. You have every sixth night off. Your brother has every tenth night off. Both of you were off on June 1. Your brother would like to see a movie with you. When will the two of you have the same night off again?

Section 5.2 The Integers; Order of Operations

- **Objective 2 - Graph integers on a number line.**

Solved Problem:	**Your Turn (exercise #2, 4, pg. 269):**
Graph:	Graph:

Graph:

    a. -4   b. 0   c. 3

Graph:

a. 5
b. -2

74

Objective 3 - Use the symbols < and >.

Solved Problem:

Insert either < or > in the underlined area between the integers to make each statement true:

a. 6 _____ -7
b. -8 _____ -1
c. -25 _____ -2
d. -14 _____ 0

a. $6 > -7$ because 6 is to the right of -7 on the number line.

b. $-8 < -1$ because -8 is to the left of -1 on the number line.

c. $-25 < -2$ because -25 is to the left of -2 on the number line.

d. $-14 < 0$ because -14 is to the left of 0 on the number line.

Your Turn (exercise #5, 7, 11, pg. 269):

Insert either < or > in the underlined area between the integers to make each statement true:

a.   -2 _____ 7
b. -13 _____ -2
c. -100 _____ 0

- **Objective 4 - Find the absolute value of an integer.**

Solved Problem:

Find the absolute value:

a. $|-8|$ b. $|6|$

a. $|-8| = 8$ because -8 is 8 units from 0.

b. $|6| = 6$ because 6 is 6 units from 0.

Your Turn (exercise #13, 15, pg. 269):

Find the absolute value:

a. $|-14|$ b. $|14|$

76

- **Objective 5 - Perform operations with integers.**

Solved Problem:

Subtract:
a. 30 - (-7)

b. -14 - (-10)

c. -14 - 10.

a. $30 - (-7) = 30 + 7 = 37$

b. $-14 - (-10) = -14 + 10 = -4$

c. $-14 - 10 = -24$

Your Turn (exercise #35, 37, 41, pg. 269):

Subtract:
a. 4 - (-10)

b. -6 - (-17)

c. -11 - 17

- **Objective 6 - Use the order of operations agreement.**

Solved Problem:

Simplify:
$(-8)^2 - (10 - 13)^2(-2)$

$(-8)^2 - (10 - 13)^2(-2)$
$= (-8)^2 - (-3)^2(-2)$
$= 64 - (9)(-2)$
$= 64 - (-18)$
$= 64 + 18$
$= 82$

Your Turn (exercise #93, pg. 269):

Simplify:
$(2 - 6)^2 - (3 - 7)^2$

78

- **Objective 2 - Reduce rational numbers.**

Solved Problem:

Reduce $\dfrac{72}{90}$ to lowest terms.

$72 = 2^3 \cdot 3^2$

$90 = 2 \cdot 5 \cdot 3^2$

Greatest Common Divisor is $2 \cdot 3^2$ or 18.

$\dfrac{72}{90} = \dfrac{72 \div 18}{90 \div 18} = \dfrac{4}{5}$

Your Turn (exercise #7, pg. 284):

Reduce $\dfrac{60}{108}$ to lowest terms.

- **Objective 3 - Convert between mixed numbers and improper fractions.**

Solved Problem:

Convert $2\dfrac{5}{8}$ to an improper fraction.

$$2\frac{5}{8} = \frac{8 \cdot 2 + 5}{8} = \frac{16 + 5}{8} = \frac{21}{8}$$

Your Turn (exercise #17, pg. 284):

Convert $12\dfrac{7}{16}$ to an improper fraction.

80

- **Objective 4 - Express rational numbers as decimals.**

Solved Problem:

Express each rational number as a decimal.

a. $\dfrac{3}{8} = 0.375$ b. $\dfrac{5}{11} = 0.\overline{45}$

a. $\dfrac{3}{8} = 0.375$

$$
\begin{array}{r}
0.375 \\
8\overline{)3.000} \\
24 \\
\hline
60 \\
56 \\
\hline
40 \\
40 \\
\hline
0
\end{array}
$$

b. $\dfrac{5}{11} = 0.\overline{45}$

$$
\begin{array}{r}
0.4545\ldots \\
11\overline{)5.0000} \\
44 \\
\hline
60 \\
55 \\
\hline
50 \\
44 \\
\hline
60 \\
55 \\
\hline
5
\end{array}
$$

Your Turn (exercise #27, 31, pg. 284):

Express each rational number as a decimal.

a. $\dfrac{7}{20}$ b. $\dfrac{9}{11}$

- **Objective 5 - Express decimals in the form $\dfrac{a}{b}$.**

Solved Problem:

Express each terminating decimal as a quotient of integers, reduced to lowest terms:

a. 0.9   b. 0.86

a. $0.9 = \dfrac{9}{10}$

b. $0.86 = \dfrac{86}{100} = \dfrac{86 \div 2}{100 \div 2} = \dfrac{43}{50}$

Your Turn (exercise #37, 43, pg. 284):

Express each terminating decimal as a quotient of integers, reduced to lowest terms:

a. 0.3   b. 0.82

82

- **Objective 6 - Multiply and divide rational numbers.**

Solved Problem:

Multiply. If possible, reduce the product to its lowest terms (from Check Point 8, parts a and c, p. 279).

a. $\dfrac{4}{11} \cdot \dfrac{2}{3}$

c. $\left(3\dfrac{2}{5}\right)\left(1\dfrac{1}{2}\right)$

a. $\dfrac{4}{11} \cdot \dfrac{2}{3} = \dfrac{8}{33}$

c. $\left(3\dfrac{2}{5}\right)\left(1\dfrac{1}{2}\right) = \dfrac{17}{5} \cdot \dfrac{3}{2} = \dfrac{51}{10}$ or $5\dfrac{1}{10}$

Your Turn (exercise #59, 63, pg. 285):

Multiply. If possible, reduce the product to its lowest terms.

a. $\left(-\dfrac{1}{10}\right)\left(\dfrac{7}{12}\right)$

b. $\left(3\dfrac{3}{4}\right)\left(1\dfrac{3}{5}\right)$

- **Objective 7 - Add and subtract rational numbers.**

Solved Problem:

Perform the indicated operations (from Check Point 10, parts b and c, p. 280):

b. $\dfrac{7}{4} - \dfrac{1}{4}$ c. $-3\dfrac{3}{8} - \left(-1\dfrac{1}{8}\right)$

b. $\dfrac{7}{4} - \dfrac{1}{4} = \dfrac{7-1}{4} = \dfrac{6}{4} = \dfrac{3}{2}$ or $1\dfrac{1}{2}$

c. $-3\dfrac{3}{8} - \left(-1\dfrac{1}{8}\right) = -\dfrac{27}{8} - \left(-\dfrac{9}{8}\right)$

$= -\dfrac{27}{8} + \dfrac{9}{8}$

$= \dfrac{-27+9}{8}$

$= \dfrac{-18}{8}$

$= -\dfrac{9}{4}$

or $-2\dfrac{1}{4}$

Your Turn (exercise #73, 95, pg. 285):

Perform the indicated operations.

a. $\dfrac{5}{6} - \dfrac{1}{6}$

b. $-1\dfrac{4}{7} - \left(-2\dfrac{5}{14}\right)$

84

- **Objective 8 - Use the order of operations agreement with rational numbers.**

Solved Problem:

Simplify:

$$\left(-\tfrac{1}{2}\right)^2 - \left(\tfrac{7}{10} - \tfrac{8}{15}\right)^2 (-18)$$

$$= \left(-\tfrac{1}{2}\right)^2 - \left(\tfrac{21}{30} - \tfrac{16}{30}\right)^2 (-18)$$
$$= \left(-\tfrac{1}{2}\right)^2 - \left(\tfrac{5}{30}\right)^2 (-18)$$
$$= \left(-\tfrac{1}{2}\right)^2 - \left(\tfrac{1}{6}\right)^2 (-18)$$
$$= \tfrac{1}{4} - \tfrac{1}{36}(-18)$$
$$= \tfrac{1}{4} + \tfrac{18}{36}$$
$$= \tfrac{1}{4} + \tfrac{2}{4}$$
$$= \tfrac{3}{4}$$

Your Turn (exercise #97, pg. 285):

Simplify:

$$\left(\frac{1}{2} - \frac{1}{3}\right) \div \frac{5}{8}$$

Objective 9 - Apply the density property of rational numbers.

Solved Problem:

Find the rational number halfway between $\frac{1}{3}$ and $\frac{1}{2}$.

First, find the sum:
$$\frac{1}{3}+\frac{1}{2} = \frac{1}{3}\cdot\frac{2}{2}+\frac{1}{2}\cdot\frac{3}{3}$$
$$= \frac{2}{6}+\frac{3}{6}$$
$$= \frac{5}{6}$$

Next, divide by 2:
$$\frac{5}{6}\div\frac{2}{1} = \frac{5}{6}\cdot\frac{1}{2}$$
$$= \frac{5}{12}$$

The rational number halfway between $\frac{1}{3}$ and $\frac{1}{2}$ is $\frac{5}{12}$.

Your Turn (exercise #107, pg. 285):

Find the rational number halfway between $\frac{1}{2}$ and $\frac{2}{3}$.

86

- **Objective 10 - Solve problems involving rational numbers.**

Solved Problem:

A chocolate-chip cookie recipe for five dozen cookies requires two eggs. If you want to make seven dozen cookies, exactly how many eggs are needed? Now round your answer to a realistic number that does not involve a fractional part of an egg.

Amount of eggs needed

$$= \frac{\text{desired serving size}}{\text{recipe serving size}} \times \text{eggs in recipe}$$

$$= \frac{7 \text{ dozen}}{5 \text{ dozen}} \times 2 \text{ eggs}$$

$$= \frac{14}{5} \text{ eggs}$$

$$= 2\frac{4}{5} \text{ eggs}$$

$$\approx 3 \text{ eggs}$$

Your Turn (exercise #129, pg. 286):

A mix for eight servings of instant potatoes requires $2\frac{2}{3}$ cups of water. If you want to make 11 servings, how much water is needed?

Section 5.4 The Irrational Numbers

- **Objective 2 - Simplify square roots.**

Solved Problem:

Simplify, if possible:

a. $\sqrt{12}$ b. $\sqrt{60}$

a. $\sqrt{12} = \sqrt{4 \cdot 3}$
$\phantom{\sqrt{12}} = \sqrt{4} \cdot \sqrt{3}$
$\phantom{\sqrt{12}} = 2\sqrt{3}$

b. $\sqrt{60} = \sqrt{4 \cdot 15}$
$\phantom{\sqrt{60}} = \sqrt{4} \cdot \sqrt{15}$
$\phantom{\sqrt{60}} = 2\sqrt{15}$

Your Turn (exercise #17, 21, pg. 296):

Simplify, if possible.

a. $\sqrt{20}$ b. $\sqrt{250}$

88

- **Objective 3 - Perform operations with square roots.**

Solved Problem:

Multiply (from Check Point 2, parts b and c, p. 291):

b. $\sqrt{10} \cdot \sqrt{10} = \sqrt{10 \cdot 10} = \sqrt{100} = 10$

b. $\sqrt{10} \cdot \sqrt{10} = \sqrt{10 \cdot 10}$
$\qquad = \sqrt{100} = 10$

c. $\sqrt{6} \cdot \sqrt{2} = \sqrt{6 \cdot 2}$
$\qquad = \sqrt{12}$
$\qquad = \sqrt{4} \cdot \sqrt{3}$
$\qquad = 2\sqrt{3}$

Your Turn (exercise #27, 29, pg. 296):

Multiply:

a. $\sqrt{6} \times \sqrt{6}$

b. $\sqrt{3} \times \sqrt{6}$

• **Objective 4 - Rationalize denominators.**

Solved Problem:

Rationalize the denominator (from Check Point 6, parts b and c, p. 294):

b. $\sqrt{\dfrac{2}{7}}$

c. $\dfrac{5}{\sqrt{18}}$

b. $\sqrt{\dfrac{2}{7}} = \dfrac{\sqrt{2}}{\sqrt{7}} = \dfrac{\sqrt{2}}{\sqrt{7}} \cdot \dfrac{\sqrt{7}}{\sqrt{7}} = \dfrac{\sqrt{14}}{\sqrt{49}} = \dfrac{\sqrt{14}}{7}$

c. $\dfrac{5}{\sqrt{18}} = \dfrac{5}{\sqrt{18}} \cdot \dfrac{\sqrt{2}}{\sqrt{2}} = \dfrac{5\sqrt{2}}{\sqrt{36}} = \dfrac{5\sqrt{2}}{6}$

Your Turn (exercise #63, 65, pg. 296):

Rationalize the denominator:

a. $\dfrac{15}{\sqrt{12}}$

b. $\sqrt{\dfrac{2}{5}}$

90

Section 5.5 Real Numbers and Their Properties; Clock Addition

- **Objective 1 - Recognize subsets of the real numbers.**

Solved Problem:

Consider the following set of numbers:

$$\left\{-9,\ -1.3,\ 0,\ 0.\overline{3},\ \frac{\pi}{2},\ \sqrt{9},\ \sqrt{10}\right\}$$

List the numbers in the set that are
a. natural numbers.
b. whole numbers.
c. integers.
d. rational numbers.
e. irrational numbers.
f. real numbers.

a. Natural numbers: $\sqrt{9}$ because $\sqrt{9} = 3$

b. Whole numbers: $0,\ \sqrt{9}$

c. Integers: $-9,\ 0,\ \sqrt{9}$

d. Rational numbers: $-9,\ -1.3,\ 0,\ 0.\overline{3},\ \sqrt{9}$

e. Irrational numbers: $\frac{\pi}{2},\ \sqrt{10}$

f. Real numbers: All numbers in this set.

Your Turn (exercise #3, pg. 308):

Consider the following set of numbers:

$$\left\{-11,\ -\tfrac{5}{6},\ 0,\ 0.75,\ \sqrt{5},\ \pi,\ \sqrt{64}\right\}$$

List the numbers in the set that are
a. natural numbers.
b. whole numbers.
c. integers.
d. rational numbers.
e. irrational numbers.
f. real numbers.

- **Objective 2 - Recognize properties of real numbers.**

Solved Problem:

Name the property illustrated (from Check Point 2, parts a, b, c, and e, p. 303).

a. $(4 \cdot 7) \cdot 3 = 4 \cdot (7 \cdot 3)$

b. $3(\sqrt{5} + 4) = 3(4 + \sqrt{5})$

c. $3(\sqrt{5} + 4) = 3\sqrt{5} + 12$

e. $1 + 0 = 1$

a. Associative property of multiplication

b. Commutative property of addition

c. Distributive property of multiplication over addition

e. Identity property of addition

Your Turn (exercise #29, 31, 35, 39, pg. 308-309):

Name the property illustrated.

a. $6 + (-4) = (-4) + 6$

b. $6 + (2 + 7) = (6 + 2) + 7$

c. $2(-8 + 6) = -16 + 12$

d. $\sqrt{17} \cdot 1 = \sqrt{17}$

- **Objective 3 - Apply properties of real numbers to clock addition.**

Solved Problem:

TABLE 5.5 4-Hour Clock Addition

$\oplus$	0	1	2	3
0	0	1	2	3
1	1	2	3	0
2	2	3	0	1
3	3	0	1	2

Your Turn (exercise #51, pg. 309):

$\oplus$	0	1	2	3	4	5	6	7
0	0	1	2	3	4	5	6	7
1	1	2	3	4	5	6	7	0
2	2	3	4	5	6	7	0	1
3	3	4	5	6	7	0	1	2
4	4	5	6	7	0	1	2	3
5	5	6	7	0	1	2	3	4
6	6	7	0	1	2	3	4	5
7	7	0	1	2	3	4	5	6

Use **Table 5.5** (p. 307) which shows clock addition in the 4-hour clock system to solve this exercise.

a. How can you tell that the set { 0, 1, 2, 3} is closed under the operation of clock addition?

b. Verify the associative property: $(2 \oplus 2) \oplus 3 = 2 \oplus (2 \oplus 2)$

c. What is the identity element in the 4-hour clock system?

d. Find the inverse of each element in the 4-hour clock system.

e. Verify two cases of the commutative property: $1 \oplus 3 = 3 \oplus 1$ and $3 \oplus 2 = 2 \oplus 3$

Shown in the figure are an 8-hour clock and the table for clock addition in the 8-hour clock system.

a. How can you tell that the set $\{0, 1, 2, 3, 4, 5, 6, 7\}$ is closed under the operation of clock addition?

b. Verify the associative property: $(4 \oplus 6) \oplus 7 = 4 \oplus (6 \oplus 7)$

c. What is the identity element in the 8-hour clock system?

d. Find the inverse of each element in the 8-hour clock system.

e. Verify two cases of the commutative property: $5 \oplus 6 = 6 \oplus 5$ and $4 \oplus 7 = 7 \oplus 4$

a. The entries in the body of the table are all elements of the set.

b. $(2 \oplus 2) \oplus 3 = 2 \oplus (2 \oplus 3)$

$0 \oplus 3 = 2 \oplus 1$

$3 = 3$

c. The identity element is 0, because it does not change anything.

d. The inverse of 0 is 0, the inverse of 1

is 3, the inverse of 2 is 2,
and the inverse of 3 is 1.

e. $1 \oplus 3 = 3 \oplus 1$ $3 \oplus 2 = 2 \oplus 3$
   $0 = 0$ $1 = 1$

Section 5.6 Exponents and Scientific Notation

- **Objective 1 - Use properties of exponents.**

Solved Problem:

Use the zero exponent rule to simplify (from Check Point 1, parts a, c, and d, p. 312):

a. 19^0 c. $(-14)^0$ d. -14^0

a. $19^0 = 1$

c. $(-14)^0 = 1$

d. $-14^0 = -1$

Your Turn (exercise #13, 15, 17, pg. 320):

Use the zero exponent rule to simplify:

a. 3^0 b. $(-3)^0$ c. -3^0

- **Objective 2 - Convert from scientific notation to decimal notation.**

Solved Problem:

Write each number in decimal notation:

a. 7.4×10^9 b. 3.017×10^{-6}

a. $7.4 \times 10^9 = 7,400,000,000$

b. $3.017 \times 10^{-6} = 0.000003017$

Your Turn (exercise #47, 55, pg. 320):

Write each number in decimal notation:

a. 8×10^7 b. 7.86×10^{-4}

96

- **Objective 3 - Convert from decimal notation to scientific notation.**

Solved Problem:

Write each number in scientific notation:

a. $7,410,000,000$

b. 0.000000092

a. $7,410,000,000 = 7.41 \times 10^9$

b. $0.000000092 = 9.2 \times 10^{-8}$

Your Turn (exercise #65, 71, pg. 320):

Write each number in scientific notation:

a. $220,000,000$

b. 0.00000293

- **Objective 4 - Perform computations using scientific notation.**

Solved Problem:

Multiply: $(1.3\times10^{7})\times(4\times10^{-2})$. Write the product in decimal notation.

$$(1.3\times10^{7})\times(4\times10^{-2}) = (1.3\times4)\times(10^{7}\times10^{-2})$$
$$= 5.2\times10^{7+(-2)}$$
$$= 5.2\times10^{5}$$
$$= 520,000$$

Your Turn (exercise #81, pg. 320):

Multiply: $\left(2\times10^{9}\right)\left(3\times10^{-5}\right)$. Write the product in decimal notation.

98

- **Objective 5 - Solve applied problems using scientific notation.**

Solved Problem:

As of December 2011, the United States had spent $2.6 trillion for the wars in Iraq and Afghanistan. (Source: costsofwar.org) At that time, the U.S. population was approximately 312 million 3.12×10^8 . If the cost of these wars was evenly divided among every individual in the United States, how much would each citizen have to pay?

$$\frac{2.6 \times 10^{12}}{3.12 \times 10^8} = \left(\frac{2.6}{3.12}\right) \times \left(\frac{10^{12}}{10^8}\right)$$

$$\approx 0.83 \times 10^4$$

$$= \$8300$$

Your Turn (exercise #111, pg. 321):

The bar graph shows the total amount Americans paid in federal taxes, in trillions of dollars, and the U.S. population, in millions, from 2007 through 2010. The exercises below are based on the numbers displayed by the graph

.

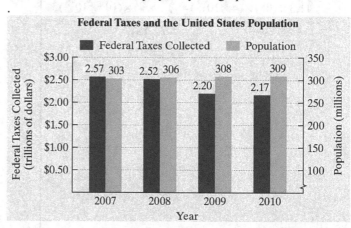

a. In 2010, the United States government collected $2.17 trillion in taxes. Express this number in scientific notation.

b. In 2010, the population of the United States was approximately 309 million. Express this number in scientific notation.

c. Use your scientific notation answers from parts (a) and (b) to answer this question: If the total 2010 tax collections were evenly divided among all Americans, how much would each citizen pay? Express the answer in scientific and decimal notations.

100

Section 5.7 Arithmetic and Geometric Sequences

- **Objective 1 - Write terms of an arithmetic sequence.**

Solved Problem:

Write the first six terms of the arithmetic sequence with first term 100 and common difference 20.

100,
100 + 20 = 120,
120 + 20 = 140,
140 + 20 = 160,
160 + 20 = 180,
180 + 20 = 200

100, 120, 140, 160, 180, and 200

Your Turn (exercise #5, pg. 329):

Write the first six terms of the arithmetic sequence with first term -7 and common difference 4.

- **Objective 2 - Use the formula for the general term of an arithmetic sequence.**

Solved Problem:

Write the first six terms of the arithmetic sequence with $a_1 = 8$ and $d = -3$.

8,
$8 - 3 = 5,$
$5 - 3 = 2,$
$2 - 3 = -1,$
$-1 - 3 = -4,$
$-4 - 3 = -7$

$8, 5, 2, -1, -4,$ and -7

Your Turn (exercise #21, pg. 329):

For the arithmetic sequence with first term a_1, find a_6 when $a_1 = 13$ and $d = 4$.

102

- **Objective 3 - Write terms of a geometric sequence.**

Solved Problem:

Write the first six terms of the geometric sequence with first term 12 and common ratio $-\dfrac{1}{2}$.

12,

$12\left(-\dfrac{1}{2}\right) = -6,$

$-6\left(-\dfrac{1}{2}\right) = 3,$

$3\left(-\dfrac{1}{2}\right) = -\dfrac{3}{2},$

$-\dfrac{3}{2}\left(-\dfrac{1}{2}\right) = \dfrac{3}{4},$

$\dfrac{3}{4}\left(-\dfrac{1}{2}\right) = -\dfrac{3}{8}$

$12, \ -6, \ 3, \ -\dfrac{3}{2}, \ \dfrac{3}{4}, \ -\dfrac{3}{8}$

Your Turn (exercise #63, pg. 329):

Write the first six terms of the geometric sequence with first term $\frac{1}{4}$ and common ratio 2.

- **Objective 4 - Use the formula for the general term of a geometric sequence.**

Solved Problem:

Find the seventh term of the geometric sequence whose first term is 5 and common ratio is -3.

$a_n = a_1 r^{n-1}$ with $a_1 = 5$, $r = -3$, and $n = 7$

$a_7 = 5(-3)^{7-1} = 5(-3)^6 = 5(729) = 3645$

Your Turn (exercise #77, pg. 329):

Find the seventh term of the geometric sequence whose first term is 5 and common ratio is -2.

Chapter 6
Algebra: Equations and Inequalities

Section 6.1 Algebraic Expressions and Formulas

- **Objective 1 - Evaluate algebraic expressions.**

Solved Problem:	**Your Turn (exercise #29, pg. 349):**
Evaluate $-3x^2 + 4xy - y^3$ for $x = 5$ and $y = -1$.	Evaluate $-x^2 + 4$ for $x = 5$.

If $x = 5$ and $y = -1$, then
$$-3x^2 + 4xy - y^3 = -3(5)^2 + 4(5)(-1) - (-1)^3$$
$$= -3(25) - 20 - (-1)$$
$$= -75 - 20 + 1$$
$$= -94$$

106

- **Objective 2 - Use mathematical models.**

Solved Problem:

The mathematical model

$$M = -120x^2 + 998x + 590$$

describes the number of calories needed per day, M, by men in age group x with moderately active lifestyles. According to the model, how many calories per day are needed by men between the ages of 19 and 30, inclusive, with this lifestyle? Does this underestimate or overestimate the number shown by the graph in **Figure 6.1** (p. 342)? By how much?

$$M = -120x^2 + 998x + 590$$
$$M = -120(4)^2 + 998(4) + 590$$
$$ = 2662$$

According to the model, men between the ages of 19 and 30 with this lifestyle need 2662 calories per day. This underestimates the actual value shown in the bar graph (2700) by 38 calories.

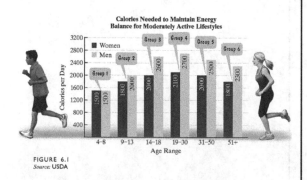

FIGURE 6.1
Source: USDA

Your Turn (exercise #69, pg. 348):

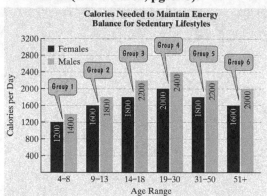

The bar graph shows the estimated number of calories per day needed to maintain energy balance for various gender and age groups for sedentary lifestyles. (Sedentary means a lifestyle that includes only the light physical activity associated with typical day-to-day life).

The mathematical model

$$F = -82x^2 + 654x + 620$$

describes the number of calories needed per day, F, by females in age group x with sedentary lifestyles. According to the model, how many calories per day are needed by females between the ages of 19 and 30, inclusive, with this lifestyle? Does this underestimate or overestimate the number shown by the graph? By how much?

- **Objective 4 - Simplify algebraic expressions.**

Solved Problem:

Simplify: $7(4x^2 + 3x) + 2(5x^2 + x)$

$$7(4x^2 + 3x) + 2(5x^2 + x) = 28x^2 + 21x + 10x^2 + 2x$$
$$= 38x^2 + 23x$$

Your Turn (exercise #57, pg. 347):

Simplify: $3(-4x^2 + 5x) - (5x - 4x^2)$

Section 6.2 Linear Equations in One Variable and Proportions

- **Objective 1 - Solve linear equations.**

Solved Problem:

Solve and check: $6(x-3)-10x = -10$.

$$6(x-3)-10x = -10$$
$$6x-18-10x = -10$$
$$-4x-18 = -10$$
$$-4x-18+18 = -10+18$$
$$-4x = 8$$
$$\frac{-4x}{-4} = \frac{8}{-4}$$
$$x = -2$$

Check:
$$6(-2-3)-10(-2) = -10$$
$$6(-5)+20 = -10$$
$$-30+20 = -10$$
$$-10 = -10$$
The solution set is $\{-2\}$.

Your Turn (exercise #19, pg. 362):

Solve and check: $14-5x = -41$

- **Objective 2 - Solve linear equations containing fractions.**

Solved Problem:

Solve and check: $\dfrac{2x}{3} = 7 - \dfrac{x}{2}$

$$\frac{2x}{3} = 7 - \frac{x}{2}$$

$$6 \cdot \frac{2x}{3} = 6 \cdot \left(7 - \frac{x}{2}\right)$$

$$6 \cdot \frac{2x}{3} = 6 \cdot 7 - 6 \cdot \frac{x}{2}$$

$$2 \cdot 2x = 42 - 3x$$

$$4x = 42 - 3x$$

$$4x + 3x = 42 - 3x + 3x$$

$$7x = 42$$

$$\frac{7x}{7} = \frac{42}{7}$$

$$x = 6$$

Check:

$$\frac{2(6)}{3} = 7 - \frac{6}{2}$$

$$\frac{12}{3} = 7 - 3$$

$$4 = 4$$

The solution set is $\{6\}$.

Your Turn (exercise #45, pg. 362):

Solve and check: $\dfrac{2x}{3} - 5 = 7$

110

- **Objective 3 - Solve proportions.**

Solved Problem:

Solve each proportion and check:

a. $\dfrac{10}{x} = \dfrac{2}{3}$ b. $\dfrac{22}{60-x} = \dfrac{2}{x}$

a. $\dfrac{10}{x} = \dfrac{2}{3}$

$10 \cdot 3 = 2x$

$30 = 2x$

$\dfrac{30}{2} = \dfrac{2x}{2}$

$15 = x$

Check:

$\dfrac{10}{15} = \dfrac{2}{3}$

$\dfrac{2}{3} = \dfrac{2}{3}$

The solution set is $\{15\}$.

b. $\dfrac{22}{60-x} = \dfrac{2}{x}$

$22x = 2(60-x)$

$22x = 120 - 2x$

$22x + 2x = 120 - 2x + 2x$

$24x = 120$

$\dfrac{24x}{24} = \dfrac{120}{24}$

$x = 5$

Check:

$\dfrac{22}{60-5} = \dfrac{2}{5}$

$\dfrac{22}{55} = \dfrac{2}{5}$

$\dfrac{2}{5} = \dfrac{2}{5}$

The solution set is $\{5\}$.

Your Turn (exercise #61, 71, pg. 362):

Solve each proportion and check:

a. $\dfrac{x}{6} = \dfrac{18}{4}$ b. $\dfrac{y+10}{10} = \dfrac{y-2}{4}$

- **Objective 4 - Solve problems using proportions.**

Solved Problem:

The property tax on a house with an assessed value of $250,000 is $3500. Determine the property tax on a house with an assessed value of $420,000, assuming the same tax rate.

Let x = the property tax on the $420,000 house.

$$\frac{\text{Tax on \$250,000 house}}{\text{Assessed value (\$250,000)}} = \frac{\text{Tax on \$420,000 house}}{\text{Assessed value (\$420,000)}}$$

$$\frac{\$3500}{\$250,000} = \frac{\$x}{\$420,000}$$

$$\frac{3500}{250,000} = \frac{x}{420,000}$$

$$250,000x = (3500)(420,000)$$

$$250,000x = 1,470,000,000$$

$$\frac{250,000x}{250,000} = \frac{1,470,000,000}{250,000}$$

$$x = 5880$$

The property tax is $5880.

Your Turn (exercise #109, pg. 363):

The volume of blood in a person's body is proportional to body weight. A person who weighs 160 pounds has approximately 5 quarts of blood. Estimate the number of quarts of blood in a person who weighs 200 pounds.

112

- **Objective 5 - Identify equations with no solution or infinitely many solutions.**

Solved Problem:

Solve: $3x + 7 = 3(x + 1)$

$$3x + 7 = 3(x + 1)$$
$$3x + 7 = 3x + 3$$
$$3x + 7 - 3x = 3x + 3 - 3x$$
$$7 = 3$$

There is no solution; the solution set is $\varnothing$, the empty set.

Your Turn (exercise #73, pg. 362):

Solve: $3x - 7 = 3(x + 1)$

Copyright © 2015 Pearson Education, Inc.

Section 6.3 Applications of Linear Equations

- **Objective 1 - Use linear equations to solve problems.**

Solved Problem:

Figure 6.5 (p. 368) shows that the freshman class of 2010 was less interested in developing a philosophy of life than the freshmen of 1969 had been. In 1969, 85% of the freshmen considered this objective essential or very important. Since then, this percentage has decreased by approximately 0.9 each year. If this trend continues, by which year will only 25% of college freshmen consider "developing a meaningful philosophy of life" essential or very important?

Let x = the number of years after 1969.
$$85 - 0.9x = 25$$
$$-0.9x = 25 - 85$$
$$-0.9x = -60$$
$$x = \frac{-60}{-0.9}$$
$$x \approx 67$$
25% of freshmen will respond this way 67 years after 1969, or 2036.

Your Turn (exercise #23, pg. 374):

Even as Americans increasingly view a college education as essential for success, many believe that a college education is becoming less available to qualified students. The exercise below is based on the data displayed by the graph.

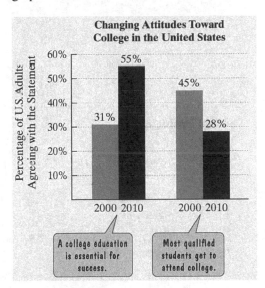

In 2000, 31% of U.S. adults viewed a college education as essential for success. For the period 2000 through 2010, the percentage viewing a college education as essential for success increased on average by approximately 2.4 each year. If this trend continues, by which year will 67% of all American adults view a college education as essential for success?

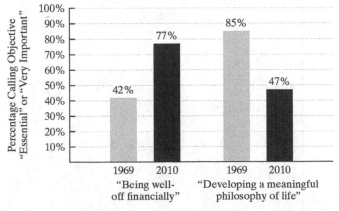

FIGURE 6.5
Source: Higher Education Research Institute

114

- **Objective 2 - Solve a formula for a variable.**

Solved Problem:

Solve the formula $T = D + pm$ for m.

$$T = D + pm$$
$$T - D = D - D + pm$$
$$T - D = pm$$
$$\frac{T - D}{p} = \frac{pm}{p}$$
$$\frac{T - D}{p} = m$$
$$m = \frac{T - D}{p}$$

Your Turn (exercise #55, pg. 375):

Solve the formula below for r. Do you recognize the formula? If so, what does it describe?

$$S = P + Prt$$

Section 6.4 Linear Inequalities in One Variable

- **Objective 1 - Graph subsets of real numbers on a number line.**

Solved Problem:	**Your Turn (exercise #1, 7, 9; pg. 384):**
Graph each set:	Graph each set:
a. $x < 4$ b. $x \geq -2$ c. $-4 \leq x < 1$	a. $\{x \mid x > 6\}$
a. $x < 4$	b. $\{x \mid x \leq 4\}$
b. $x \geq -2$	
c. $-4 \leq x < 1$	c. $\{x \mid -2 < x \leq 5\}$

116

- **Objective 2 - Solve linear inequalities.**

Solved Problem:

Solve and graph the solution set:
$$7x - 3 > 13x + 33$$

$$7x - 3 > 13x + 33$$
$$7x - 3 + 3 > 13x + 33 + 3$$
$$7x > 13x + 36$$
$$7x - 13x > 13x + 36 - 13x$$
$$-6x > 36$$
$$\frac{-6x}{-6} < \frac{36}{-6}$$
$$x < -6$$
$$\{x \mid x < -6\}$$

Your Turn (exercise #49, pg. 384):

Solve and graph the solution set:

$$2x - 5 < 5x - 11$$

- **Objective 3 - Solve applied problems using linear inequalities.**

Solved Problem:

To earn a B in a course, you must have a final average of at least 80%. On the first three examinations, you have grades of 82%, 74%, and 78%. If the final examination counts as two grades, what must you get on the final to earn a B in the course?

Let x = your grade on the final exam.

$$\frac{82+74+78+x+x}{5} \geq 80$$

$$\frac{234+2x}{5} \geq 80$$

$$5\left(\frac{234+2x}{5}\right) \geq 5(80)$$

$$234+2x \geq 400$$

$$234+2x-234 \geq 400-234$$

$$2x \geq 166$$

$$\frac{2x}{2} \geq \frac{166}{2}$$

$$x \geq 83$$

You need at least an 83% on the final to get a B in the course.

Your Turn (exercise #89, pg. 385):

A car can be rented from Continental Rental for $80 per week plus 25 cents for each mile driven. How many miles can you travel if you can spend at most $400 for the week?

118

- **Objective 1 - Multiply binomials using the FOIL method.**

Solved Problem:

Multiply: $(7x+5)(4x-3)$

$$(7x+5)(4x-3) = 7x \cdot 4x + 7x(-3) + 5 \cdot 4x + 5(-3)$$
$$= 28x^2 - 21x + 20x - 15$$
$$= 28x^2 - x - 15$$

Your Turn (exercise #5, pg. 399):

Multiply: $(2x-1)(x+2)$

- **Objective 2 - Factor trinomials.**

Solved Problem:

Factor: $5x^2 - 14x + 8$

$5x^2 - 14x + 8 = (5x - 4)(x - 2)$

Your Turn (exercise #27, pg. 399):

Factor: $3x^2 - 25x - 28$

120

- **Objective 3 - Solve quadratic equations by factoring.**

Solved Problem:

Solve: $2x^2 + 7x - 4 = 0$

$$2x^2 + 7x - 4 = 0$$
$$(2x - 1)(x + 4) = 0$$

$$2x - 1 = 0 \quad \text{or} \quad x + 4 = 0$$
$$2x = 1 \qquad\qquad x = -4$$
$$x = \frac{1}{2}$$

The solution set is $\left\{ -4, \ \frac{1}{2} \right\}$

Your Turn (exercise #39, pg. 399):

Solve: $x^2 - 2x - 15 = 0$

- **Objective 4 - Solve quadratic equations using the quadratic formula.**

Solved Problem:

Solve using the quadratic formula:
$$2x^2 = 6x - 1$$

$$2x^2 = 6x - 1$$

$$2x^2 - 6x + 1 = 0$$

$$a = 2, \quad b = -6, \quad c = 1$$

$$x = \frac{-b \pm \sqrt{b^2 - 4ac}}{2a}$$

$$x = \frac{-(-6) \pm \sqrt{(-6)^2 - 4(2)(1)}}{2(2)}$$

$$x = \frac{6 \pm \sqrt{36 - 8}}{4}$$

$$x = \frac{6 \pm \sqrt{28}}{4}$$

$$x = \frac{6 \pm 2\sqrt{7}}{4}$$

$$x = \frac{2\left(3 \pm \sqrt{7}\right)}{4}$$

$$x = \frac{3 \pm \sqrt{7}}{2}$$

$$x = \frac{3 + \sqrt{7}}{2} \quad \text{or} \quad x = \frac{3 - \sqrt{7}}{2}$$

The solution set is $\left\{ \dfrac{3 + \sqrt{7}}{2}, \ \dfrac{3 - \sqrt{7}}{2} \right\}$

Your Turn (exercise #69, pg. 400):

Solve using the quadratic formula:

$$6x^2 + 6x + 1 = 0$$

122

- **Objective 5 - Solve problems modeled by quadratic equations.**

Solved Problem:

The formula $P = 0.01A^2 + 0.05A + 107$ models a woman's normal systolic blood pressure, P, at age A. Use this formula to find the age, to the nearest year, of a woman whose normal systolic blood pressure is 115 mm Hg. Use the blue graph in **Figure 6.6** (p. 397) to verify your solution.

$$P = 0.01A^2 + 0.05A + 107$$

$$115 = 0.01A^2 + 0.05A + 107$$

$$0 = 0.01A^2 + 0.05A - 8$$
$$a = 0.01, \quad b = 0.05, \quad c = -8$$

$$A = \frac{-b \pm \sqrt{b^2 - 4ac}}{2a}$$

$$A = \frac{-(0.05) \pm \sqrt{(0.05)^2 - 4(0.01)(-8)}}{2(0.01)}$$

$$A = \frac{-0.05 \pm \sqrt{0.3225}}{0.02}$$

$$A \approx \frac{-0.05 + \sqrt{0.3225}}{0.02} \qquad A \approx \frac{-0.05 - \sqrt{0.3225}}{0.02}$$

$$A \approx 26 \qquad\qquad A \approx -31$$

Age cannot be negative, reject the negative answer. Thus, a woman whose normal systolic blood pressure is 115 mm Hg is 26 years old.

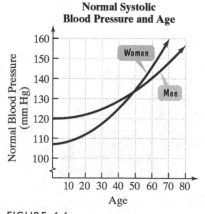

Normal Systolic Blood Pressure and Age

FIGURE 6.6

Your Turn (exercise #85a, pg. 400):

The percentage, p, of the United States population that was foreign-born x years after 1920 can be modeled by the formula

$$p = 0.004x^2 - 0.36x + 14$$

According to the model, what percentage of the U.S. population was foreign-born in 2000? Does the model underestimate the actual number displayed by the bar graph? By how much?

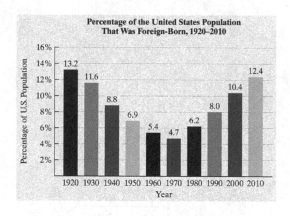

Chapter 7
Algebra: Graphs, Functions, and Linear Systems

Section 7.1 Graphing and Functions

- **Objective 1 – Plot points in the rectangular coordinate system.**

Solved Problem:

Plot the points: $A(-3, 5)$, $B(2, -4)$, $C(5, 0)$, $D(-5, -3)$, $E(0, 4)$, and $F(0, 0)$.

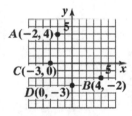

Your Turn (exercise #1, 5, 9, 11, pg. 418):

Plot the given points in a rectangular coordinate system.

a. $(1, 4)$

b. $(-3, -5)$

c. $(-4, 0)$

d. $(0, -3)$

124

- **Objective 2 - Graph equations in the rectangular coordinate system.**

Solved Problem:

Graph $y = 4 - x$. Select integers for x, starting with -3 and ending with 3.

x	$y = 4 - x$	(x, y)
-3	$y = 4 - (-3) = 4 + 3 = 7$	$(-3, 7)$
-2	$y = 4 - (-2) = 4 + 2 = 6$	$(-2, 6)$
-1	$y = 4 - (-1) = 4 + 1 = 5$	$(-1, 5)$
0	$y = 4 - (0) = 4 - 0 = 4$	$(0, 4)$
1	$y = 4 - (1) = 4 - 1 = 3$	$(1, 3)$
2	$y = 4 - (2) = 4 - 2 = 2$	$(2, 2)$
3	$y = 4 - (3) = 4 - 3 = 1$	$(3, 1)$

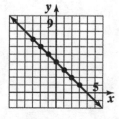

Your Turn (exercise #23, pg. 418):

Graph $y = x - 2$. Select integers for x from -3 to 3, inclusive.

- **Objective 3 - Use function notation.**

Solved Problem:

Find each of the following:

a. $f(6)$ for $f(x) = 4x + 5$

b. $g(-5)$ for $g(x) = 3x^2 - 10$

c. $h(-4)$ for $h(r) = r^2 - 7r + 2$.

a. $f(x) = 4x + 5$

$f(6) = 4(6) + 5$

$= 29$

b. $g(x) = 3x^2 - 10$

$g(-5) = 3(-5)^2 - 10$

$= 65$

c. $h(r) = r^2 - 7r + 2$

$h(-4) = (-4)^2 - 7(-4) + 2$

$= 46$

Your Turn (exercise #35a, 41b, 43b, pg. 418):

Evaluate each function at the given value of the variable.

a. $f(x) = 3x - 2$ for $f(7)$

b. $h(r) = 3r^2 + 5$ for $h(-1)$

c. $f(x) = 2x^2 + 3x - 1$ for $f(-4)$

126

- **Objective 4 - Graph functions.**

Solved Problem:

Graph the functions $f(x) = 2x$ and $g(x) = 2x - 3$ in the same rectangular coordinate system. Select integers for x from -2 to 2, inclusive. How is the graph of g related to the graph of f?

x	$f(x) = 2x$	(x, y) or $(x, f(x))$
-2	$f(-2) = 2(-2) = -4$	$(-2, -4)$
-1	$f(-1) = 2(-1) = -2$	$(-1, -2)$
0	$f(0) = 2(0) = 0$	$(0, 0)$
1	$f(1) = 2(1) = 2$	$(1, 2)$
2	$f(2) = 2(2) = 4$	$(2, 4)$

x	$g(x) = 2x - 3$	(x, y) or $(x, f(x))$
-2	$g(-2) = 2(-2) - 3 = -7$	$(-2, -7)$
-1	$g(-1) = 2(-1) - 3 = -5$	$(-1, -5)$
0	$g(0) = 2(0) - 3 = -3$	$(0, -3)$
1	$g(1) = 2(1) - 3 = -1$	$(1, -1)$
2	$g(2) = 2(2) - 3 = 1$	$(2, 1)$

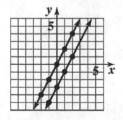

Your Turn (exercise #47, pg. 418):

Evaluate $f(x)$ for the given values of x. Then use the ordered pairs $(x, f(x))$ from your table to graph the function.

$$f(x) = x^2 - 1$$

x	$f(x) = x^2 - 1$

- **Objective 5 - Use the vertical line test.**

Solved Problem:

Use the vertical line test to identify graphs in which *y* is a function of *x*. (From Check Point 7a and 7c on p. 415).

a.

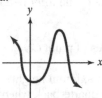

b.

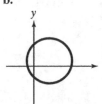

a. *y* is a function of *x*.

c. *y* is not a function of *x*. Two values of *y* correspond to an *x*-value

Your Turn (exercise #59, 61, pg. 418-419):

Use the vertical line text to identify graphs in which *y* is a function of *x*.

a.

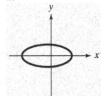

b.

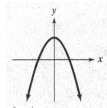

128

- **Objective 6 -Obtain information about a function from its graph.**

Solved Problem:

When a person receives a drug injected into a muscle, the concentration of the drug in the body, measured in milligrams per 100 milliliters, depends on the time elapsed after the injection, measured in hours. **Figure 7.15** (p. 417) shows the graph of drug concentration over time, where x represents hours after the injection and y represents the drug concentration at time x.

a. During which period of time is the drug concentration increasing?
b. During which period of time is the drug concentration decreasing?
c. What is the drug's maximum concentration and when does this occur?
d. What happens by the end of 13 hours?
e. Explain why the graph defines y as a function of x.

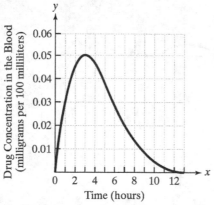

FIGURE 7.15

a. The concentration is increasing from 0 to 3 hours.
b. The concentration is decreasing from 3 to 13
hours.
c. The maximum concentration of 0.05 mg per
100 ml occurs after 3 hours.
d. None of the drug is left in the body.

Your Turn (exercise #71, 73, 75, pg. 419)

A football is thrown by a quarterback to a receiver. The points in the figure show the height of the football, in feet, above the ground in terms of its distance, in yards, from the quarterback.

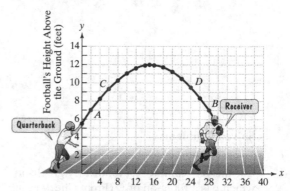

Distance of the Football from the Quarterback (yards)

a. Find the coordinates of point A. Then interpret the coordinates in terms of the information given.

b. Estimate the coordinates of point C.

c. What is the football's maximum height? What is its distance from the quarterback when it reaches its maximum height?

e. The graph defines y as a function of x because
no vertical line intersects the graph in more than one point.

Section 7.2 Linear Functions and Their Graphs

- **Objective 1 - Use intercepts to graph a linear equation.**

Solved Problem:

Graph: $2x + 3y = 6$.

Find the x-intercept by setting $y = 0$
$2x + 3(0) = 6$

$2x = 6$

$x = 3$; resulting point $(3, 0)$

Find the y-intercept by setting $x = 0$
$2(0) + 3y = 6$

$3y = 6$

$y = 2$; resulting point $(0, 2)$

Find a checkpoint by substituting any value.
$2(1) + 3y = 6$

$2 + 3y = 6$

$3y = 4$

$y = \dfrac{4}{3}$; resulting point $\left(1, \dfrac{4}{3}\right)$

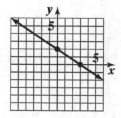

Your Turn (exercise #3, pg. 430):

Use the x- and y-intercepts to graph

$3x - 4y = 12$

- **Objective 2 - Calculate slope.**

Solved Problem:

Find the slope of the line passing through each pair of points:

a. (-3, 4) and (-4, -2)
b. (4, -2) and (-1, 5).

a. $m = \dfrac{-2-4}{-4-(-3)} = \dfrac{-6}{-1} = 6$

b. $m = \dfrac{5-(-2)}{-1-4} = \dfrac{7}{-5} = -\dfrac{7}{5}$

Your Turn (exercise #11, 13, pg. 430):

Calculate the slope of the line passing through the given points. If the slope is undefined, so state. Then indicated if whether the line rises, falls, is horizontal, or is vertical.

a. $(-2,1)$ and $(3,5)$

b. $(-2,4)$ and $(-1,-1)$

132

- **Objective 3 - Use the slope and y-intercept to graph a line.**

Solved Problem:

Graph the linear function $y = \dfrac{3}{5}x + 1$ by using the slope and y-intercept.

Step 1. Plot the y-intercept of $(0, 1)$

Step 2. Obtain a second point using the slope m.

$m = \dfrac{3}{5} = \dfrac{\text{Rise}}{\text{Run}}$

Starting from the y-intercept move up 3 units and move 5 units to the right. This puts the second point at $(5, 4)$.

Step 3. Draw the line through the two points.

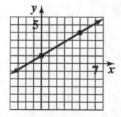

Your Turn (exercise #25, pg. 430):

Graph the linear function $y = \dfrac{1}{2}x + 3$ using the slope and using the slope and y-intercept.

- **Objective 4 - Graph horizontal or vertical lines.**

Solved Problem:

Graph $y = 3$ in the rectangular coordinate system.

Draw a horizontal line that intersects the y-axis at 3.

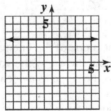

Your Turn (exercise #43, pg. 430):

Graph $y = -2$

134

- **Objective 5 - Interpret slope as rate of change.**

Solved Problem:

Find the slope of the line segment representing men in **Figure 7.25**. (p. 428) Use your answer to complete this statement:

For the period from 1970 through 2010, the percentage of married men ages 20 to 24 decreased by _____ per year. The rate of change is _____ per _____.

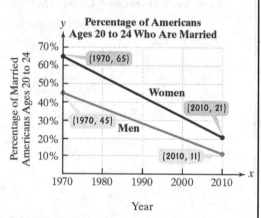

FIGURE 7.25
Source: U.S. Census Bureau

The two points shown on the line segment for Medicare are (2007, 446) and (2016, 909).

$$m = \frac{11-45}{2010-1970} = \frac{-34}{40} = -0.85$$

For the period from 1970 through 2010, the percentage of married men ages 20 to 24 decreased by 0.85 per year. The rate of change −0.85% is per year.

Your Turn (exercise #59, pg. 431):

As we age, daily stress and worry decrease and happiness increases, according to an analysis of 340,847 U.S. adults, ages 18-85, in the journal *Proceedings of the National Academy of Sciences.* The graphs show a portion of the research.

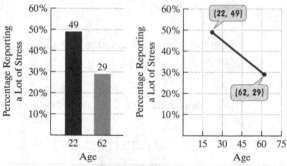

a. Find the slope of the line passing through the two points shown by the voice balloons. Express the slope as a decimal.

b. Use your answer from part (a) to complete the statement:

For each year of aging, the percentage of Americans reporting "a lot" of stress decreases by _____ %. The rate of change is _____ % per _____.

- **Objective 6 - Use slope and y-intercept to model data**

Solved Problem:

a. Use the two points for college in **Figure 7.26(b)** (p. 429) to find a function in the form $C(x) = mx + b$ that models the percentage of college graduates in the U.S. population, $C(x)$, x years after 1960.

b. Use the model to project the percentage of college graduates in 2020.

a. The y-intercept is 8 and the slope is $m = \dfrac{\text{Change in } y}{\text{Change in } x} = \dfrac{24 - 8}{50 - 0} = \dfrac{16}{50} = 0.32$

The equation is $C(x) = 0.32x + 8$.

b. $C(x) = 0.32x + 8$

$C(60) = 0.32(60) + 8$

$ = 27.2$

The model projects that 27.2% of the U.S. population will be college graduates in 2020.

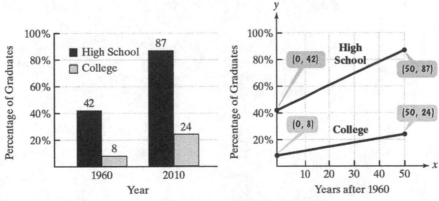

Percentage of High School Graduates and College Graduates in the U.S. Population

FIGURE 7.26(a)   FIGURE 7.26(b)

Source: James M. Henslin, *Essentials of Sociology*, Ninth Edition, Pearson, 2011.

Your Turn (exercise #63, pg. 432):

Big (Lack of) Men on Campus. The bar graph shows the number of bachelor's degrees, in thousands, awarded to men and women in the United States for four selected years from 1980 to 2010. The trend indicated by the graphs is among the hottest topics of debate among college-admissions officers. Some private liberal arts colleges have quietly begun special efforts to recruit men – including admissions preferences for them.

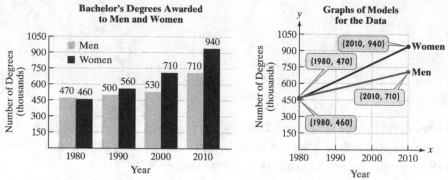

a. Use the two points for women shown by the blue voice balloons to find a function in the form $W(x) = mx + b$ that models the number of bachelor's degrees, $W(x)$, in thousands, awarded to women x years after 1980.

b. Use the model from part (a) to project the number of bachelor's degrees that will be awarded to women in 2020.

Section 7.3 Systems of Linear Equations in Two Variables

- **Objective 1 - Decide whether an ordered pair is a solution of a linear system.**

Solved Problem:

Determine whether (-4, 3) is a solution of the system:

$$\begin{cases} x + 2y = 2 \\ x - 27 = 6 \end{cases}$$

Replace x with -4 and y with 3.

$$x + 2y = 2 \qquad\qquad x - 2y = 6$$
$$-4 + 2(3) = 2 \qquad -4 - 2(3) = 6$$
$$-4 + 6 = 2 \qquad\qquad -4 - 6 = 6$$
$$2 = 2 \text{ true} \qquad -10 = 6 \text{ false}$$

The pair $(-4, 3)$ does not satisfy both equations. Therefore it is not a solution of the system.

Your Turn (exercise #1, pg. 444):

Determine whether (2,3) is a solution of the system:

$$\begin{cases} x + 3y = 11 \\ x - 5y = -13 \end{cases}$$

138

- **Objective 2 - Solve linear systems by graphing.**

Solved Problem:

Solve by graphing:

$$\begin{cases} 2x + 3y = 6 \\ 2x + y = -2 \end{cases}$$

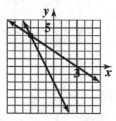

Check coordinates of intersection:

$$2x + 3y = 6 \qquad\qquad 2x + y = -2$$
$$2(-3) + 3(4) = 6 \qquad 2(-3) + (4) = -2$$
$$-6 + 12 = 6 \qquad\qquad -6 + 2 = -2$$
$$6 = 6, \text{ true} \qquad\qquad -2 = -2, \text{ true}$$

The solution set is $\{(-3, 4)\}$.

Your Turn (exercise #7, pg. 444):

Solve by graphing:

$$\begin{cases} 2x - 3y = 6 \\ 4x + 3y = 12 \end{cases}$$

- **Objective 3 - Solve linear systems by substitution.**

Solved Problem:

Solve by the substitution method:

$$\begin{cases} 3x + 2y = -1 \\ x - y = 3 \end{cases}$$

Step 1. Solve one of the equations for one variable:

$x - y = 3$

$\quad x = y + 3$

Step 2. Substitute into the other equation:

$\quad\quad 3x + 2y = -1$

$\quad\quad\quad\quad \overset{x}{\overbrace{\quad\quad}}$

$3(y + 3) + 2y = -1$

Step 3. Solve: $3(y + 3) + 2y = -1$

$\quad\quad 3y + 9 + 2y = -1$

$\quad\quad\quad 5y + 9 = -1$

$\quad\quad\quad\quad 5y = -10$

$\quad\quad\quad\quad\quad y = -2$

Step 4. Back-substitute the obtained value into the equation from step 1:

$x = y + 3$

$x = -2 + 3$

$x = 1$

Step 5. Check $(1, -2)$ in both equations:

$\quad x - y = 3 \quad\quad\quad\quad 3x + 2y = -1$

$1 - (-2) = 3 \quad\quad 3(1) + 2(-2) = -1$

$\quad\quad 3 = 3, \text{ true} \quad\quad\quad -1 = -1, \text{ true}$

The solution set is $\{(1, -2)\}$.

Your Turn (exercise #17, pg. 444):

Solve by the substitution method:

$$\begin{cases} x + 3y = 5 \\ 4x + 5y = 13 \end{cases}$$

140

Objective 4 - Solve linear systems by addition.

Solved Problem:

Solve by the addition method:
$$\begin{cases} 3x = 2 - 4y \\ 5y = -1 - 2x \end{cases}$$

Rewrite both equations in the form
$Ax + By = C$:

$3x = 2 - 4y \quad \rightarrow \quad 3x + 4y = 2$

$5y = -1 - 2x \quad \rightarrow \quad 2x + 5y = -1$

Rewrite with opposite coefficients, then add and solve:

$3x + 4y = \;\; 2 \xrightarrow{\text{Mult. by 2}} \quad 6x + 8y = 4$

$2x + 5y = -1 \xrightarrow{\text{Mult. by -3}} \quad \underline{-6x - 15y = 3}$

$\qquad\qquad\qquad\qquad\qquad\qquad -7y = 7$

$\qquad\qquad\qquad\qquad\qquad\qquad\quad y = -1$

Back-substitute into either equation:

$3x = 2 - 4y$

$3x = 2 - 4(-1)$

$3x = 6$

$\;x = 2$

Checking confirms the solution set is $\{(2, -1)\}$.

Your Turn (exercise #35, pg. 444):

Solve by the addition method:

$$\begin{cases} 2x = 3y - 4 \\ -6x + 12y = 6 \end{cases}$$

- **Objective 5 - Identify systems that do not have exactly one ordered-pair solution.**

Solved Problem:

Solve the system:
$$\begin{cases} x + 2y = 4 \\ 3x + 6y = 13 \end{cases}$$

Rewrite with a pair of opposite coefficients, then add:

$x + 2y = 4 \xrightarrow{\text{Mult. by } -3} -3x - 6y = -12$

$3x + 6y = 13 \xrightarrow{\text{No change}} \underline{3x + 6y = 13}$

$\phantom{3x + 6y = 13 \xrightarrow{\text{No change}}} 0 = 1$

The statement $0 = 1$ is false which indicates that the system has no solution. The solution set is the empty set, $\varnothing$.

Your Turn (exercise #37, pg. 445):

Solve the system:

$$\begin{cases} x = 9 - 2y \\ x + 2y = 13 \end{cases}$$

142

• **Objective 6 - Solve problems using systems of linear equations.**

Solved Problem:

A company that manufactures running shoes has a fixed cost of $300,000. Additionally, it costs $30 to produce each pair of shoes. They are sold at $80 per pair.

a. Write the cost function, C, of producing x pairs of running shoes.
b. Write the revenue function, R, from the sale of x pairs of running shoes.
c. Determine the break-even point. Describe what this means.

a. $C(x) = 300,000 + 30x$

b. $R(x) = 80x$

c. $R(x) = C(x)$
$80x = 300,000 + 30x$
$50x = 300,000$
$x = 6000$

$C(6000) = 300,000 + 30(6000) = 480,000$

Break even point (6000, 480000)
The company will need to make 6000 pairs of shoes and earn $480,000 to break even.

Your Turn (exercise #57, pg. 445):

A company that manufactures small canoes has a fixed cost of $18,000. It costs $20 to produce each canoe. The selling price is $80 per canoe. (In solving this exercise, let x represent the number of canoes that canoes produced and sold.)

a. Write the cost function, C.

b. Write the revenue function, R.

c. Determine the break-even point. Describe what this means.

Section 7.4 Linear Inequalities in Two Variables

- **Objective 1 - Graph a linear inequality in two variables.**

Solved Problem:

Graph: $2x - 4y \geq 8$

Graph the equation $2x - 4y = 8$ as a solid line. Choose a test point that is not on the line.

$$\underline{\text{Test } (0,0)}$$
$$2x - 4y \geq 8$$
$$2(0) - 4(0) \geq 8$$
$$0 \geq 8, \text{ false}$$

Since the statement is false, shade the other half-plane.

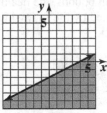

Your Turn (exercise #7, pg. 454):

Graph: $5x + 3y \leq -15$

144

- **Objective 2 - Use mathematical models involving linear inequalities.**

Solved Problem:

Show that point B in **Figure 7.38** (p. 451) is a solution of the system of inequalities that describes healthy weight.

Point $B = (66, 130)$
$$4.9x - y \geq 165$$
$$4.9(66) - 130 \geq 165$$
$$193.4 \geq 165, \text{ true}$$

$$3.7x - y \leq 125$$
$$3.7(66) - 130 \leq 125$$
$$114.2 \leq 125, \text{ true}$$

Point B is a solution of the system.

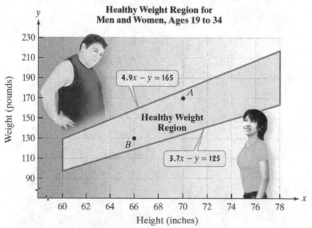

FIGURE 7.38
Source: U.S. Department of Health and Human Services

Your Turn (exercise #45, pg. 454):

The figure shows the healthy weight region for various heights for people ages 35 and older.

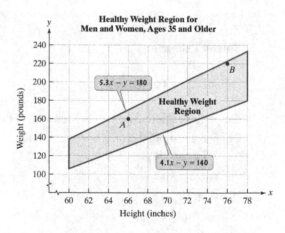

If x represents height, in inches, and y represents weight, in pounds, the healthy weight region can be modeled by the following system of linear inequalities:

$$\begin{cases} 5.3x - y \geq 180 \\ 4.1x - y \leq 140 \end{cases}$$

Show that point A is a solution of the system of inequalities that described healthy weight for this age group.

- **Objective 3 - Graph a system of linear inequalities.**

Solved Problem:

Graph the solution set of the system:

$$\begin{cases} x + 2y > 4 \\ 2x - 3y \le -6 \end{cases}$$

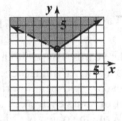

Graph the equation $x + 2y = 4$ as a dashed line.
Choose a test point that is not on the line.

Test $(0,0)$

$x + 2y > 4$

$0 + 2(0) > 4$

$0 > 4$, false

Since the statement is false, shade the other half-plane.

Next, graph the equation $2x - 3y = -6$ as a solid line.
Choose a test point that is not on the line.

Test $(0,0)$

$2x - 3y \le -6$

$2(0) - 3(0) \le -6$

$0 \le -6$, false

Since the statement is false, shade the other half-plane.

The graph is the intersection (overlapping) of the two half-planes.

Your Turn (exercise #27, pg. 454):

Graph the solution set of the system:

$$\begin{cases} 2x + y < 4 \\ x - y > 4 \end{cases}$$

146

- **Objective 1 - Write an objective function describing a quantity that must be maximized or minimized.**

Solved Problem:

From Check Point 1 on page 457:
A company manufactures bookshelves and desks for computers. Let x represent the number of bookshelves manufactured daily and y the number of desks manufactured daily. The company's profits are $25 per bookshelf and $55 per desk. Write the objective function that describes the company's total daily profit, z, from x bookshelves and y desks. (Check Points 2 through 4 are also related to this situation, so keep track of your answers.)

The total profit is 25 times the number of bookshelves, x, plus 55 times the number of desks, y. The objective function is
$z = 25x + 55y$

Your Turn (exercise #13a, pg. 461):

A student earns $10 per hour tutoring and $7 per hour as a teacher's aide. Let x = the number of hours each week spent tutoring and y = the number of hours each week spent as a teacher's aide. Write the objective function that describes the total weekly earnings.

- **Objective 2 - Use inequalities to describe limitations in a situation.**

Solved Problem:

A company manufactures bookshelves and desks for computers. Let x represent the number of bookshelves manufactured daily and y the number of desks manufactured daily. The company's profits are $25 per bookshelf and $55 per desk.

From Check Point 2 on page 457:
To maintain high quality, the company in Check Point 1 should not manufacture more than a total of 80 bookshelves and desks per day. Write an inequality that describes this constraint.

Not more than a total of 80 bookshelves and desks can be manufactured per day. This is represented by the inequality $x + y \leq 80$.

Your Turn (exercise #13b, pg. 461):

A student earns $10 per hour tutoring and $7 per hour as a teacher's aide. Let x = the number of hours each week spent tutoring and y = the number of hours each week spent as a teacher's aide. The student is bound by these constraints:

- To have enough time for studies, the student can work no more than 20 hours per week

- The tutoring center requires that each tutor spend at least three hours per week tutoring

- The tutoring center requires that each tutor spend no more than eight hours per week tutoring.

Write a system of three inequalities that describes these constraints.

148

- **Objective 3 - Use linear programming to solve problems.**

Solved Problem:

A company manufactures bookshelves and desks for computers. Let x represent the number of bookshelves manufactured daily and y the number of desks manufactured daily. The company's profits are $25 per bookshelf and $55 per desk.

To meet customer demand, the company must manufacture between 30 and 80 bookshelves per day, inclusive. Furthermore, the company must manufacture at least 10 and no more than 30 desks per day. Write an inequality that describes each of these sentences. Then summarize what you have described about this company by writing the objective function for its profits and the three constraints.

Objective function: $z = 25x + 55y$

Constraints: $x + y \le 80$

$\qquad 30 \le x \le 80$

$\qquad 10 \le y \le 30$

How many bookshelves and how many desks should be manufactured by this company per day to obtain a maximum profit? What is the maximum daily profit?

Graph the constraints and find the corners, or vertices, of the region of intersection.

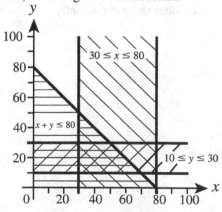

Your Turn (exercise #13c, d, e, pg. 461):

A student earns $10 per hour tutoring and $7 per hour as a teacher's aide. Let x = the number of hours each week spent tutoring and y = the number of hours each week spent as a teacher's aide. The student is bound by these constraints:

- To have enough time for studies, the student can work no more than 20 hours per week

- The tutoring center requires that each tutor spend at least three hours per week tutoring

- The tutoring center requires that each tutor spend no more than eight hours per week tutoring.

a. Graph the system of inequalities that describes these constraints. Use only the first quadrant and its boundary because x and y are nonnegative.

b. Evaluate the objective function for total weekly earnings at each of the four vertices of the graphed region. [The vertices should occur at (3,0), (8, 0), (3,17), and (8,12).]

c. Complete the missing portions of this statement: The student can earn the maximum amount per week by tutoring for ___ hours per week and working as a teacher's aide for ___ hours per week. The maximum amount that the student can earn each week is $___.

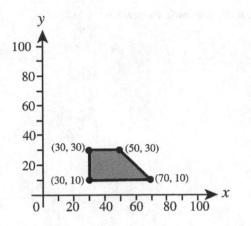

Find the value of the objective function at each corner of the graphed region.

Corner (x, y)	Objective Function $z = 25x + 55y$
(30, 10)	$z = 25(30) + 55(10)$ $= 750 + 550 = 1300$
(30, 30)	$z = 25(30) + 55(30)$ $= 750 + 1650 = 2400$
(50, 30)	$z = 25(50) + 55(30)$ $= 1250 + 1650 = 2900 \leftarrow$ Maximum
(70, 10)	$z = 25(70) + 55(10)$ $= 1750 + 550 = 2300$

The maximum value of z is 2900 and it occurs at the point (50, 30).

In order to maximize profit, 50 bookshelves and 30 desks must be produced each day for a profit of $2900.

150

- **Objective 1 - Graph exponential functions.**

Solved Problem:

Graph: $f(x) = 3^x$

x	$f(x) = 3^x$
-2	$\dfrac{1}{9}$
-1	$\dfrac{1}{3}$
0	1
1	3
2	9

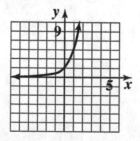

Your Turn (exercise #1, pg. 474):

Use a table of coordinates to graph the exponential function $f(x) = 4^x$. Being by selecting -2, -1, 0, 1, and 2 for x.

- **Objective 2 - Use exponential models.**

Solved Problem:

Medical research indicates that the risk of having a car accident increases exponentially as the concentration of alcohol in the blood increases. The risk is modeled by

$$R = 6e^{12.77x},$$

where x is the blood-alcohol concentration and R, given as a percent, is the risk of having a car accident. In every state, it is illegal to drive with a blood-alcohol concentration of 0.08 or greater.

In many states, it is illegal for drivers under 21 years old to drive with a blood-alcohol concentration of 0.01 or greater. What is the risk of a car accident with a blood-alcohol concentration of 0.01? Round to one decimal place.

$$R = 6e^{12.77x}$$
$$= 6e^{12.77(0.01)}$$
$$= 6.8\%$$

The risk of a car accident with a blood alcohol concentration of 0.01 is 6.8%.

Your Turn (exercise #33b, pg. 476):

Average annual premiums for employer-sponsored family health insurance policies more than doubled over 11 years. The bar graph shows the average cost of a family health insurance plan in the United States for six selected years from 2000 through 2011.

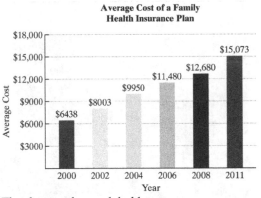

Average Cost of a Family Health Insurance Plan

The data can be modeled by
$$f(x) = 782x + 6564 \text{ and}$$

$$g(x) = 6875e^{0.077x}, \text{ in which } f(x) \text{ and}$$

$g(x)$ represent the average cost of a family health insurance plan x years after 2000.

According to the exponential model, what was the average cost of a family health insurance plan in 2011?

152

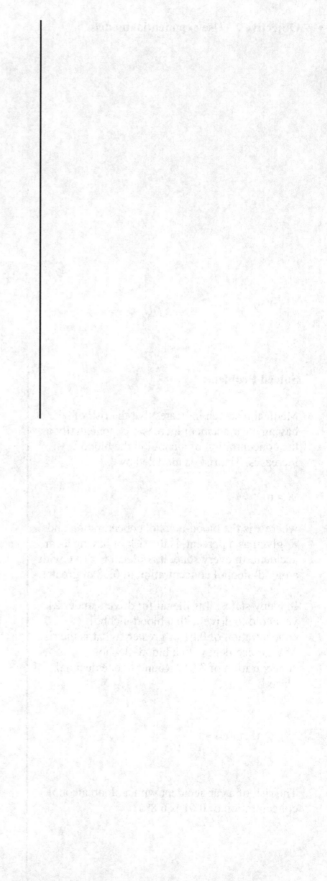

- **Objective 3 - Graph logarithmic functions.**

Solved Problem:

Rewrite $y = \log_3 x$ in exponential form. Then use the exponential form of the equation to obtain the function's graph. Select integers from -2 to 2, inclusive, for y.

$y = \log_3 x$ is equivalent to $x = 3^y$.

$x = 3^y$	y	(x, y)
$\frac{1}{9}$	-2	$\left(\frac{1}{9}, -2\right)$
$\frac{1}{3}$	-1	$\left(\frac{1}{3}, -1\right)$
1	0	$(1, 0)$
3	1	$(3, 1)$
9	2	$(9, 2)$

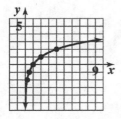

Your Turn (exercise #7, pg. 474):

Rewrite $y = \log_4 x$ in exponential form. Then use a table of coordinates and the exponential form to graph the logarithmic function. Begin by selecting -2, -1, 0, 1, and 2 for y.

154

- **Objective 4 - Use logarithmic models.**

Solved Problem:

Use the model obtained in Example 5(a)

$$f(x) = -11.6 + 13.4 \ln x$$

to find the temperature increase, to the nearest degree, after $x = 30$ minutes. How well does the function model the actual increase shown in **Figure 7.50(a)** (p. 469)?

Temperature Increase in an Enclosed Vehicle

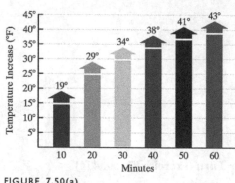

FIGURE 7.50(a)
Source: Professor Jan Null, San Francisco State University

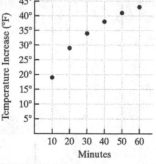

FIGURE 7.50(b)

$$f(x) = -11.6 + 13.4 \ln x$$
$$f(30) = -11.6 + 13.4 \ln 30$$
$$f(30) \approx 34°$$

The function models the actual data extremely well.

Your Turn (exercise #37a, pg. 477):

The percentage of adult height attained by a girl who is x years old can be modeled by

$$f(x) = 62 + 35 \log(x - 4)$$

where x represents the girl's age (from 5 to 15) and $f(x)$ represents the percentage of adult height.

According to the model, what percentage of her adult height has a girl attained at age 13? Use a calculator and round to the nearest tenth of a percent.

- **Objective 5 - Graph quadratic functions.**

Solved Problem:

Graph the quadratic function:
$y = x^2 + 6x + 5$

Step 1. Since $a > 0$, the parabola opens upward $(a = 1)$.

Step 2. Find the vertex given $a = 1$ and $b = 6$.

x-coordinate of vertex $= \dfrac{-b}{2a} = \dfrac{-6}{2(1)} = \dfrac{-6}{2} = -3$

y-coordinate of vertex
$= (-3)^2 + 6(-3) + 5 = 9 - 18 + 5 = -4$
Thus, the vertex is the point $(-3, -4)$.

Step 3.   Replace y with 0 and solve the equation for x by factoring.

$x^2 + 6x + 5 = 0$

$(x + 5)(x + 1) = 0$

$x + 5 = 0 \quad$ or $\quad x + 1 = 0$

$x = -5 \qquad\qquad x = -1$

Thus the x-intercepts are -5 and -1, , which are located at the points $(-5, 0)$ and $(-1, 0)$.

Step 4. Replace x with 0 and solve the equation for y.

$y = x^2 + 6x + 5$

$y = (0)^2 + 6(0) + 5$

$y = 5$

Thus the y-intercept is 5, which is located at the point (0, 5).

Steps 5 and 6. Plot the intercepts and the vertex. Connect these points with a smooth curve.

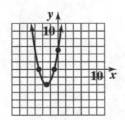

Your Turn (exercise #39c, pg. 477):

A ball is thrown upward and outward from a height of 6 feet. The table shows four measurements indicating the ball's height at various horizontal distances from where it was thrown. The graphing calculator screen displays a quadratic function that models the ball's height, y, in feet, in terms of its horizontal distance, x, in feet.

x, Ball's Horizontal Distance (feet)	y, Ball's Height (feet)
0	6
1	7.6
3	6
4	2.8

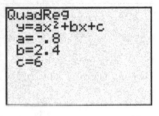

Use the graphing calculator screen to express the model in function notation.

156

Objective 6 - Use quadratic models.

Solved Problem:

Use the model obtained in Example 7(a) (p. 472)

$$f(x) = -0.01x^2 + 1.18x + 2$$

(where x is a football's horizontal distance in feet from the defensive player) to answer this question:

If the defensive player had been 8 feet from the kicker's point of impact, how far would he have needed to reach to block the punt? Does this seem realistic? Identify the solution as a point on the graph in **Figure 7.58** (p. 473).

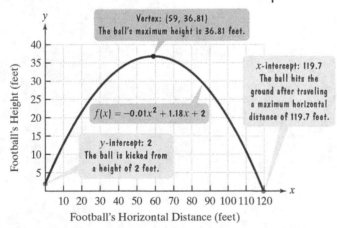

FIGURE 7.58 The parabolic path of a punted football

$$f(x) = -0.01x^2 + 1.18x + 2$$
$$f(8) = -0.01(8)^2 + 1.18(8) + 2$$
$$= 10.8$$

The defensive player would have needed to reach 10.8 feet. It would be unlikely any player could reach that far. This would be represented on the graph by the point $(8, 10.8)$.

Your Turn:

Use the model from the previous "Your Turn" exercise to determine the x-coordinate of the quadratic function's vertex. Then complete this statement: The maximum height of the ball occurs ____ feet from where it was thrown and the maximum height is ____ feet.

158

Chapter 8
Personal Finance

Section 8.1 Percent, Sales Tax, and Discount

- **Objective 1 - Express a fraction as a percent.**

Solved Problem:

Express $\dfrac{1}{8}$ as a percent.

Step 1: $\dfrac{1}{8} = 1 \div 8 = 0.125$

Step 2: $0.125 \cdot 100 = 12.5$

Step 3: 12.5%

Your Turn (exercise #1, pg. 495):

Express 2/5 as a percent.

160

- **Objective 2 - Express a decimal as a percent.**

Solved Problem:

Express 0.023 as a percent.

$$0.023 = 2.3 \%$$

Your Turn (exercise #13, pg. 495):

Express 0.3844 as a percent.

- **Objective 3 - Express a percent as a decimal.**

Solved Problem:

Express each percent as a decimal.

a. 67%
b. 250%

a. 67% = 0.67

b. 250% = 2.50

Your Turn (exercises #21, 25, pg. 495):

Express each percent as a decimal.

72%
130%

162

- **Objective 4 - Solve applied problems involving sales tax and discounts.**

Solved Problem:

Suppose that the local sales tax is 6% and you purchase a computer for $1260.

a) How much tax is paid?
b) What is the computer's total cost?

a) 6% of $1260 = $0.06 \times \$1260 = \$75.60$
The tax paid is $75.60

b) $\$1260.00 + \$75.60 = \$1335.60$
The total cost is $1335.60

Your Turn (exercise #47, pg. 495):

Suppose that the local sales tax rate is 6% and you purchase a car for $32,800.

a) How much tax is paid?
b) What is the car's total cost?

- **Objective 5 - Determine percent increase or decrease.**

Solved Problem:

A television regularly sells for $940. The sale price is $611. Find the percentage decrease of the sale price from the regular price.

Amount of decrease: $940 − $611 = $329

$$\frac{\text{amount of decrease}}{\text{original amount}} = \frac{\$329}{\$940} = 0.35 = 35\%$$

There was a 35% decrease in price.

Your Turn (exercise #55, pg. 496):

A sofa regularly sells for $840. The sale price is $714. Find the percent decrease of the sale price from the regular price.

164

- **Objective 6 - Investigate some of the ways percent can be abused.**

Solved Problem:

Suppose you paid $1200 in taxes. During year 1, taxes decrease by 20%. During year 2, taxes increase by 20%.

a) What do you pay in taxes for year 2?
b) How do your taxes for year 2 compare with what you originally paid, namely $1200? If the taxes are not the same, find the percent increase or decrease.

a. 20% of $1200 $= 0.20 \times \$1200 = \$240$

Taxes for year 1 are $\$1200 - \$240 = \$960$

20% of $960 $= 0.20 \times \$960 = \$192$

Taxes for year 2 are $\$960 + \$192 = \$1152$

b. $\dfrac{\$1200 - \$1152}{\$1200} = \dfrac{\$48}{\$1200} = 0.04 = 4\%$

Taxes for year 2 are 4% less than the original amount.

Your Turn (exercise #57, pg. 496):

Suppose that you have $10,000 in a rather risky investment recommended by your financial advisor. During the first year, your investment decreases by 30% of its original value. During th second year, your investment increases by 40% of its first-year value. Your advisor tells you that there must have been a 10% overall increase of your original $10,000 investment. Is your financial advisor using percentages properly? If not, what is your actual percent gain or loss of your original $10,000 investment?

Section 8.2 Income Tax

- **Objective 1 - Determine gross income, adjustable gross income, and taxable income.**

Solved Problem:

A single woman earned wages of $87,200, received $2680 in interest from a savings account, and contributed $3200 to a tax-deferred savings plan. She is entitled to a personal exemption of $3800 and a standard deduction of $5950. The interest on her home mortgage was $11,700, she paid $4300 in property taxes and $5220 in state taxes, and she contributed $15,000 to charity.

a) Determine the woman's gross income.
b) Determine the woman's adjusted gross income.
c) Determine the woman's taxable income.

Your Turn (exercise #1, pg. 506):

A taxpayer earned wages of $52,600, received $720 in interest from a savings account, and contributed $3200 to a tax-deferred retirement plan. He was entitled to a personal exemption of $3800 and had deductions totaling $5450.

a) Determine the man's gross income
b) Determine the man's adjusted gross income.
c) Determine the man's taxable income.

a. Gross income $= \$87,200 + \$2680 = \$89,880$

b. Adjusted gross income $= \overbrace{\$89,880}^{\text{Gross income}} - \overbrace{\$3200}^{\text{Contribution}}$

$\qquad\qquad\qquad\quad = \$86,680$

c. Itemized deductions $= \overbrace{\$11,700}^{\text{Interest}} + \overbrace{\$4300}^{\substack{\text{Property}\\\text{tax}}} + \overbrace{\$5220}^{\substack{\text{State}\\\text{tax}}} + \overbrace{\$15,000}^{\text{Charity}}$

$\qquad\qquad\qquad\quad = \$36,220$

This person is entitled to a standard deduction of $5950. However, the itemized deduction is greater than the standard deduction.

Taxable income $= \overbrace{\$86,680}^{\text{Adj. gross income}} - \overbrace{(\$3800 + \$36,220)}^{\text{Exemptions and deductions}}$

$\qquad\qquad\quad = \$86,680 - \$40,020$

$\qquad\qquad\quad = \$46,660$

166

- **Objective 2 - Calculate federal income tax.**

Solved Problem:

Use the 2012 marginal tax rates in **Table 8.1** on page 501 to calculate the federal tax owed by a single mom with no dependents whose gross income, adjustments, deductions, and credits are given as follows:

Gross income: $40,000
Adjustments: $1000
Deductions: $3000: charitable contributions
$1500: theft loss
$300: cost of tax preparation
Tax credit: none

Step 1. Determine the adjusted gross income.

$$\text{Adjusted gross income} = \overbrace{\$40,000}^{\text{Gross income}} - \overbrace{\$1000}^{\text{Adjustments}}$$
$$= \$39,000$$

Step 2. Determine the taxable income.
Since the total deduction of $4800 is less than the standard deduction of $5950, use $5950.

$$\text{Taxable income} = \overbrace{\$39,000}^{\text{Adj. gross income}} - \overbrace{(\$3800 + \$5950)}^{\text{Exemptions and deductions}}$$
$$= \$39,000 - \$9750$$
$$= \$29,250$$

Step 3. Determine the income tax.

$$\text{Tax Computation} = 0.10(8700) + 0.15(29,250 - 8700)$$
$$= \$3952.50$$

$$\text{Income tax} = \overbrace{\$3952.50}^{\text{Tax Computation}} - \overbrace{0}^{\text{Tax credits}}$$
$$= \$3952.50$$

Your Turn (exercise #15, pg. 506):

Use the 2012 marginal tax rates in **Table 8.1** on page 501 to calculate the income tax owed by a single male with no dependents, whose gross income, adjustments, deductions, and credits are given as follows:

Gross income: $75,000
Adjustments: $4000
Deductions: $28,000 mortgage
interest
$4200 property taxes
$3000 charitable
contributions
Tax credit: none

- **Objective 3 - Calculate FICA taxes.**

Solved Problem:

If you are not self-employed and earn $200,000, what are your FICA taxes?

FICA Tax (not self-employed):

$$\text{FICA Tax} = \overbrace{0.0565 \times \$110,000}^{\substack{5.65\% \text{ on first} \\ \$110,000}} + \overbrace{0.0145 \times (\$200,000 - \$110,000)}^{\substack{1.45\% \text{ on income in} \\ \text{excess of } \$110,000}}$$

$$= 0.0565 \times \$110,000 + 0.0145 \times \$90,000$$

$$= \$6215 + \$1305$$

$$= \$7520$$

Your Turn (exercise #21, pg. 507):

If you are self-employed and earn $150,000, what are your FICA taxes?

168

Objective 4 - Solve problems involving working students and taxes.

Solved Problem:

You decide to work part-time at a local nursery. The job pays $12 per hour and you work 15 hours per week. Your employer withholds 10% of your gross pay for federal taxes, 5.65% for FICA taxes, and 4% for state taxes.

a. What is your weekly gross pay?
b. How much is withheld per week for federal taxes?
c. How much is withheld per week for FICA taxes?
d. How much is withheld per week for state taxes?
e. What is your weekly net pay?
f. What percentage of your gross pay is withheld for taxes? Round to the nearest tenth of a percent.

a. Gross pay $= 15$ hours $\times \dfrac{\$12}{\text{hour}} = \$180$

b. Federal taxes $= 10\% \times \$180 = 0.10 \times \$180 = \$18$

c. FICA taxes $= 5.65\% \times \$180 = 0.0565 \times \$180 = \$10.17$

d. State taxes $= 4\% \times \$180 = 0.04 \times \$180 = \$7.20$

e. Net pay $= \overbrace{\$180}^{\text{Gross pay}} - \overbrace{(\$18 + \$10.17 + \$7.20)}^{\text{Federal, FICA, and state taxes}}$

$\qquad = \$180 - \$35.37$

$\qquad = \$144.63$

f. Percent of gross pay withheld for taxes

$\quad = \dfrac{\text{Taxes}}{\text{Gross pay}} = \dfrac{\$35.37}{\$180} = 0.1965 \approx 19.7\%$

Your Turn (exercise #25, pg. 507):

You decide to work part-time at a local supermarket. The job pays $8.50 per hour and you work 20 hours per week. Your employer withholds 10% of your gross pay for federal taxes, 5.65% for FICA taxes, and 3% for state taxes.

a. What is your weekly gross pay?
b. How much is withheld per week for federal taxes?
c. How much is withheld per week for FICA taxes?
d. How much is withheld per week for state taxes?
e. What is your weekly net pay?
f. What percentage of your gross pay is withheld for taxes? Round to the nearest tenth of a percent.

Section 8.3 Simple Interest

- **Objective 1 - Calculate simple interest.**

Solved Problem:

A student took out a simple interest loan for $2400 for two years at a rate of 7%. What is the interest on the loan?

$$I = Prt = (\$2400)(0.07)(2) = \$336$$

Your Turn (exercise #3, pg. 512):

The principal P is borrowed at simple interest rate r for a period of time t. Find the simple interest owed for the use of the money. Assume 360 days in a year.

$P = \$180$, $r = 3\%$, $t = 2$ years

170

- **Objective 2 - Use the future value formula.**

Solved Problem:

A loan of $2040 has been made at 7.5% for four months. Find the loan's future value.

$$A = P(1+rt) = 2040\left[1+(0.075)\left(\frac{4}{12}\right)\right] = \$2091$$

Your Turn (exercise #13, pg. 512):

The principal P is borrowed at simple interest rate r for a period of time t. Find the loan's future value, A, or the total amount due at time t.

$P = \$24,000$, $r = 8.5\%$, $t = 8$ months

- **Objective 1 - Use compound interest formulas.**

Solved Problem:

A sum of $10,000 is invested at an annual rate of 8%. Find the balance in the account after five years subject to **a.** quarterly compounding and **b.** continuous compounding.

a. $A = P\left(1 + \dfrac{r}{n}\right)^{nt}$

$A = 10,000\left(1 + \dfrac{0.08}{4}\right)^{4(5)}$

$\qquad = \$14,859.47$

b. $A = Pe^{rt}$

$A = 10,000e^{0.08(5)}$

$\qquad = \$14,918.25$

Your Turn (exercise #15, pg. 521):

Suppose that you have $12,000 to invest. Which investment yields the greater return over three years: 7% compounded monthly or 6.85% compounded continuously?

172

- **Objective 2 - Calculate present value.**

Solved Problem:

How much money should be deposited today in an account that earns 7% compounded weekly so that it will accumulate to $10,000 in eight years?

$$P = \frac{A}{\left(1 + \dfrac{r}{n}\right)^{nt}}$$

$A = \$10,000, \; r = 0.07, \; n = 52, \; t = 8$

$$P = \frac{10,000}{\left(1 + \dfrac{0.07}{52}\right)^{52 \cdot 8}} \approx \frac{10,000}{1.750013343} \approx \$5714.25$$

Your Turn (exercise #19, pg. 521):

How much money should be deposited today in an account that earns 9.5% compounded monthly so that it will accumulate to $10,000 in three years?

- **Objective 3 - Understand and compute effective annual yield.**

Solved Problem:

What is the effective annual yield of an account paying 8% compounded quarterly?

$$Y = \left(1 + \frac{r}{n}\right)^n - 1$$

$$Y = \left(1 + \frac{0.08}{4}\right)^4 - 1 \approx 0.0824 = 8.24\%$$

Your Turn (exercise #25, pg. 521):

What is the effective annual yield of a savings account paying 6% monthly?

Section 8.5 Annuities, Methods of Saving, and Investments

- **Objective 1 - Determine the value of an annuity.**

Solved Problem:

At age 30, to save for retirement, you decide to deposit $100 at the end of each month into an IRA that pays 9.5% compounded monthly.

a. How much will you have from the IRA when you retire at age 65?

b. Find the interest.

Round answers to the nearest dollar.

a. $A = \dfrac{P\left[\left(1+\frac{r}{n}\right)^{nt}-1\right]}{\frac{r}{n}}$

$A = \dfrac{100\left[\left(1+\frac{0.095}{12}\right)^{12\times35}-1\right]}{\frac{0.095}{12}}$

$\approx \$333,946$

b. $\$333,946 - \$100 \cdot 12 \cdot 35 = \$291,946$

Your Turn (exercise #5, pg. 536):

Periodic Deposit = $50 at the end of each month
Rate = 6% compounded monthly
Time = 30 years

a. Find the value the annuity. Round to the nearest dollar.

b. Find the interest.

- **Objective 2 - Determine regular annuity payments needed to achieve a financial goal.**

Solved Problem:

Parents of a baby girl are in a financial position to begin saving for her college education. They plan to have $100,000 in a college fund in 18 years by making regular, end-of-month deposits in an annuity that pays 9% compounded monthly.

a. How much should they deposit each month? Round up to the nearest dollar.

b. How much of the $100,000 college fund comes from deposits and how much comes from interest?

a.
$$P = \frac{A\left(\frac{r}{n}\right)}{\left[\left(1+\frac{r}{n}\right)^{nt} - 1\right]}$$

$$P = \frac{100,000\left(\frac{0.09}{12}\right)}{\left[\left(1+\frac{0.09}{12}\right)^{12\times18} - 1\right]}$$

$$\approx \$187$$

b. Deposits: $\$187 \times 18 \times 12 = \$40,392$

Interest: $\$100,000 - \$40,392 = \$59,608$

Your Turn (exercise #15, pg. 536):

Periodic Deposit = $? At the end of each month
Rate = 7.25% compounded monthly
Time = 40 years
Financial Goal = $1,000,000

a. Determine the periodic deposit. Round up to the nearest dollar.

b. How much of the financial goal comes from deposits and how much comes from the interest?

176

- **Objective 4 - Read stock tables.**

Solved Problem:

Use the stock table for Coca-Cola on page 533 to solve parts (a) through (h) in Example 5 for Coca-Cola .

a. High price = $63.38, Low price = $42.37

b. Dividend = $0.72 · 3000 = $2160

c. Annual return for dividends alone = 1.5%
1.5% is much lower than the 3.5% bank rate.

d. Shares traded = 72,032 · 100 = 7,203,200 shares

e. High price = $49.94, Low price = $48.33

f. Price at close = $49.50

g. The price went up $0.03 per share

h. Annual earnings per share $= \dfrac{\$49.50}{37} \approx \$1.34$

Your Turn (exercise #19, pg. 536):

Use the stock tables for Goodyear (the tire company) on page 536.

a. What were the high and low prices for a share for the past 52 weeks?

b. If you owned 700 shares of this stock last year, what dividend did you receive?

c. What is the annual return for the dividends along? How does this compare to a bank offering a 3% interest rate?

d. How many shares of this company's stock were traded yesterday?

e. What were the high and low prices for a share yesterday?

f. What was the price at which a share last traded when the stock exchange closed yesterday?

g. What was the change in price for a share of stock from the market close two days ago to yesterday's market close?

h. Compute the company's annual earnings per share using Annual earnings per share = Yesterday's Closing price per share
        PE Ratio

- **Objective 5 - Understand accounts designed for retirement savings.**

Solved Problem:

a. Suppose that between the ages of 25 and 40, you contribute $2000 per year to a 401(k) and your employer contributes $1000 per year on your behalf. The interest rate is 8% compounded annually. What is the value of the 401(k), rounded to the nearest dollar, after 15 years?

b. After 15 years of working for this firm, you move on to a new job. However, you keep your accumulated retirement funds in the 401(k). How much money, to the nearest dollar, will you have in the plan when you reach age 65?

c. What is the difference between the amount of money you will have accumulated in the 401(k) and the amount you contributed to the plan?

a. $A = \dfrac{P\left[(1+r)^t - 1\right]}{r}$

$A = \dfrac{3000\left[(1+0.08)^{15} - 1\right]}{0.08}$

$\approx \$81,456$

b. $A = P(1+r)^t$

$A = \$81,456(1+0.08)^{25}$

$\approx \$557,849$

c. $\$557,849 - \$2000(15) = \$527,849$

Your Turn (exercise #35, pg. 538):

a. Suppose that between the ages of 22 and 40, you contribute $3000 per year to a 401(k) and your employer contributes $1500 per year on your behalf. The interest rate is 8.3% compounded annually. What is the value of the 401(k), rounded to the nearest dollar, after 18 years?

b. Suppose that after 18 years of working for this firm, you move on to a new job. However, you keep your accumulated retirement funds in the 401(k). How much money, to the nearest dollar, will you have in the plan when you reach age 65?

c. What is the difference between the amount of money you will have accumulated in the 401(k), and the amount you contributed to the plan?

- **Objective 1 - Compute the monthly payment and interest costs for a car loan.**

Solved Problem:

Suppose that you decide to borrow $15,000 for a new car. You can select one of the following loans, each requiring regular monthly payments:

Installment Loan A: four-year loan at 8%
Installment Loan B: six-year loan at 10%.

a. Find the monthly payments and the total interest for Loan A.

b. Find the monthly payments and the total interest for Loan B.

c. Compare the monthly payments and total interest for the two loans.

a.
$$PMT = \frac{P\left(\frac{r}{n}\right)}{1-\left(1+\frac{r}{n}\right)^{-nt}} = \frac{15,000\left(\frac{0.08}{12}\right)}{1-\left(1+\frac{0.08}{12}\right)^{-12(4)}} \approx \$366$$

Total interest for loan A:
$\$366 \cdot 12 \cdot 4 - \$15,000 = \$2568$

b.
$$PMT = \frac{P\left(\frac{r}{n}\right)}{1-\left(1+\frac{r}{n}\right)^{-nt}} = \frac{15,000\left(\frac{0.10}{12}\right)}{1-\left(1+\frac{0.10}{12}\right)^{-12(6)}} \approx \$278$$

Total interest for loan B:
$\$278 \cdot 12 \cdot 6 - \$15,000 = \$5016$

c. Monthly payments are lower with the longer-term loan, but there is more interest with the longer-term loan.

Your Turn (exercise #3, pg. 546):

Suppose that you decide to borrow $15,000 for a new car. You can select one of the following loans, each requiring regular monthly payments:

Installment Loan A: three-year loan at 5.1%
Installment Loan B: five-year loan at 6.4%

a. Find the monthly payments and the total interest for Loan A.

b. Find the monthly payments and the total interest for Loan B.

c. Compare the monthly payments and the total interest for the two loans.

- **Objective 5 - Compare monthly payments on new and used cars.**

Solved Problem:

Suppose that you are thinking about buying a car and have narrowed down your choices to two options:

The new-car option: The new cars costs $19,000 and can be financed with a three-year loan at 6.18%.

The used-car option: A two-year-old model of the same car costs %11,500 and can be financed with a three-year loan at 7.5%

What is the difference in monthly payments between financing the new car and financing the used car?

New car:

$$PMT = \frac{P\left(\frac{r}{n}\right)}{1-\left(1+\frac{r}{n}\right)^{-nt}} = \frac{19,000\left(\frac{0.0618}{12}\right)}{1-\left(1+\frac{0.0618}{12}\right)^{-12(3)}} \approx \$580$$

Used car:

$$PMT = \frac{P\left(\frac{r}{n}\right)}{1-\left(1+\frac{r}{n}\right)^{-nt}} = \frac{11,500\left(\frac{0.075}{12}\right)}{1-\left(1+\frac{0.075}{12}\right)^{-12(3)}} \approx \$358$$

The difference is $\$580 - \$358 = \$222$

Your Turn (exercise #5, pg. 546):

Suppose that you are thinking about buying a car and have narrowed down your choices to two options:

The new-car option: The new car costs $28,000 and can be financed with a four-year loan at 6.12%

The used-car option: A Three-year old model of the same car costs $16,000 and can be financed with a four-year loan at 6.86%.

What is the difference in monthly payments between financing the new car and financing the used car?

- **Objective 6 - Solve problems related to owning and operating a car.**

Solved Problem:

Suppose that you drive 36,000 miles per year and gas averages $3.50 per gallon.

a. What will you save in annual fuel expenses by owning a hybrid car averaging 40 miles per gallon rather than an SUV averaging 15 miles per gallon?

b. If you deposit your monthly fuel savings at the end of the month into an annuity that pays 7.25% compounded monthly, how much will you have saved at the end of the seven years?

Round all computations to the nearest dollar.

a. Annual fuel expense for the hybrid

$$= \frac{\text{annual miles driven}}{\text{miles per gallon}} \times \text{price per gallon}$$

$$= \frac{36,000}{40} \times \$3.50$$

$$= \$3150$$

Annual fuel expense for the SUV

$$= \frac{\text{annual miles driven}}{\text{miles per gallon}} \times \text{price per gallon}$$

$$= \frac{36,000}{15} \times \$3.50$$

$$= \$8400$$

The difference is $\$8400 - \$3150 = \$5250$

b. Monthly savings $= \dfrac{\$5250}{12} \approx \$438$

Thus, $A = \dfrac{P\left[\left(1 + \frac{r}{n}\right)^{nt} - 1\right]}{\frac{r}{n}}$

$$A = \frac{438\left[\left(1 + \frac{0.0725}{12}\right)^{12 \times 7} - 1\right]}{\frac{0.0725}{12}}$$

$$\approx \$47,746$$

Your Turn (exercise #11, pg. 547):

Suppose that you drive 40,000 miles per year and gas averages $4 per gallon.

a. What will you save in annual fuel expenses by owning a hybrid car averaging 40 miles per gallon rather than an SUV average 16 miles per gallon?

b. If you deposit your monthly fuel savings at the end of each month into an annuity that pays 5.2% compounded monthly, how much will you have saved at the end of six years?

Section 8.7 The Cost of Home Ownership

- **Objective 1 - Compute the monthly payment and interest costs for a mortgage.**

Solved Problem:

In Example 1 on pages 549 - 550, the $175,000 mortgage was financed with a 30-year fixed rate at 7.5%. The total interest paid over 30 years was approximately $266,220.

a. Use the loan payments formula for installment loans to find the monthly payment if the time of the mortgage is reduced to 15 years. Round to the nearest dollar.

b. Find the total interest paid over 15 years.

a. $PMT = \dfrac{P\left(\frac{r}{n}\right)}{1-\left(1+\frac{r}{n}\right)^{-nt}} = \dfrac{175,500\left(\frac{0.075}{12}\right)}{1-\left(1+\frac{0.075}{12}\right)^{-12\cdot15}} \approx \$1627$

b. $\$1627 \cdot 12 \cdot 15 - \$175,500 = \$117,360$

Your Turn (exercise #1 d, e, pg. 555-556):

The price of a home is $220,000. The bank requires a 20% down payment and three points at the time of closing. The cost of the home is financed with a 30-year fixed-rate mortgage at 7%.

d. Find the monthly payment (excluding escrowed taxes and insurance)

e. Find the total cost of interest over 30 years.

182

- **Objective 2 - Prepare a partial loan amortization schedule.**

Solved Problem:

Prepare a loan amortization schedule for the first two months of the mortgage loan shown. Round entries to the nearest cent.

Annual % Rate: 7.0%
Amount of Mortgage: $200,000
Number of Monthly Payments: 240
Monthly Payment: $1550.00
Term: Years 20

Interest for first month =

$$Prt = \$200,000 \times 0.07 \times \frac{1}{12} \approx \$1166.67$$

Principle payment =
$1550.00 - $1166.67 = $383.33

Balance of loan =
$200,000 - $383.33 = $199,616.67

Interest for second month =

$$Prt = \$199,616.67 \times 0.07 \times \frac{1}{12} \approx \$1164.43$$

Principle payment =
$1550.00 - $1164.43 = $385.57

Balance of loan =
$199,616.67 - $385.57 = $199,231.10

Payment Number	Interest Payment	Principal Payment	Balance of Loan
1	$1166.67	$383.33	$199,616.67
2	$1164.43	$385.57	$199,231.10

Your Turn (exercise #9b, pg. 556):

The cost of a home is financed with a $120,000 30-year fixed-rate mortgage at 4.5%.

b. Prepare a loan amortization schedule for the first three months of the mortgage. Round entries to the nearest cent.

Payment Number	Interest	Principal	Loan Balance
1			
2			
3			

- **Objective 3 - Solve problems involving what you can afford to spend for a mortgage.**

Solved Problem:	**Your Turn (exercise #11, pg. 556):**
Suppose that your gross annual income is $240,000.	Suppose that your gross annual income is $36,000.
a. What is the maximum amount you should spend each month on a mortgage payment?	a. What is the maximum amount you should spend each month on a mortgage payment?
b. What is the maximum amount you should spend each month for total credit obligations?	b. What is the maximum amount you should spend each month for total credit obligations?
c. If your monthly mortgage payment is 90% of the maximum amount you can afford, what is the maximum amount you should spend each month for all other debt?	c. If your monthly mortgage payment is 70% of the maximum you can afford, what is the maximum amount you should spend each month for all other debt?
Round all computations to the nearest dollar.	

Monthly gross income $= \dfrac{\$240,000}{12} = \$20,000$

a. You should spend no more than 28% on a mortgage payment:

$$28\% \times \$20,000 = 0.28 \times \$20,000 = \$5600$$

b. You should spend no more than 36% on total monthly debt:

$$36\% \times \$20,000 = 0.36 \times \$20,000 = \$7200$$

c. $\$7200 - \$5600(0.90) = \$2160$

184

- **Objective 1 - Find the interest, the balance due, and the minimum monthly payment for credit card loans.**

Solved Problem:

A credit card company calculates interest using the average daily balance method. The monthly interest rate is 1.6% of the average daily balance. The following transactions occurred during the May 1–May 31 billing period. Answer parts (a) through (d) in Example 1 on page 558 using this information.

Transaction Description	Transaction Amount
Previous balance, $8240.00	
May 1 Billing date	
May 7 Payment	$ 350.00 credit
May 15 Charge: Computer	$ 1405.00
May 17 Charge: Restaurant	$ 45.20
May 30 Charge: Clothing	$ 180.72
May 31 End of billing period	
Payment Due Date: June 9	

a. Make a table that shows the unpaid balance for each transaction date, the number of days at each unpaid balance, and then multiply each unpaid balance by the number of days that the balance was outstanding.

Date	Unpaid Balance	Number of Days at Each Unpaid Balance	$\left(\begin{array}{c}\text{Unpaid}\\\text{Balance}\end{array}\right)\cdot\left(\begin{array}{c}\text{Number}\\\text{of Days}\end{array}\right)$
May 1	$8240.00	6	$49,440.00
May 7	$8240.00 − $350.00 = $7890.00	8	$63,120.00
May 15	$7890.00 + $1405.00 = $9295.00	2	$18,590.00
May 17	$9295.00 + $45.20 = $9340.20	13	$121,422.60
May 30	$9340.20 + $180.72 = $9520.92	2	$19,041.84
		Total days: 31	Total: $271,614.44

$$\text{Average daily balance} = \frac{\text{Sum of unpaid balances}}{\text{Number of days in the billing period}}$$

$$= \frac{\$271,614.44}{31}$$

$$\approx \$8,761.76$$

b. $I = \Pr t$

$= (\$8761.76)(0.016)(1)$

$\approx \$140.19$

c. Balance due $= \$9520.92 + \$140.19 = \$9661.11$

d. Because the balance exceeds $360, the minimum payment is $\frac{1}{36}$ of the balance due.

$$\text{Minimum Payment} = \frac{\$9661.11}{36} \approx \$269$$

Your Turn (exercise #1, pg 564)

This exercise involves credit cards that calculate interest using the average daily balance method. The monthly interest rate is 1.5% of the average daily balance. Each exercise shows transactions that occurred during the March 1 – March 31 billing period.

Transaction Period	Transaction Amount
Previous balance, $6240.00	
March 1 Billing Date	
March 5 Payment	$300 credit
March 7 Charge: Restaurant	$40
March 12 Charge: Groceries	$90
March 21 Charge: Car Repairs	$230
March 31 End of billing period	
Payment Due Date: April 9	

a. Find the average daily balance for the billing period. Round to the nearest cent.

b. Find the interest to be paid on April 1, the next billing date. Round to the nearest cent.

c. Find the balance due on April 1.

d. This credit card requires a $10 minimum monthly payment if the balance due at the end of the billing period is less that $360. Otherwise, the minimum monthly payment is 1/36 of the balance due at the end of the billing period, rounded up to the nearest whole dollar. What is the minimum monthly payment due by April 9?

186

Chapter 9
Measurement

Section 9.1 **Measuring Length; The Metric System**

- **Objective 1 - Use dimensional analysis to change units of measurement.**

Solved Problem:

Convert:

a. 78 inches to feet
b. 17,160 feet to miles
c. 3 inches to yards.

a. 78 in. $= \dfrac{78 \text{ in.}}{1} \cdot \dfrac{1 \text{ ft}}{12 \text{ in.}} = 6.5 \text{ ft}$

b. 17,160 ft $= \dfrac{17,160 \text{ ft}}{1} \cdot \dfrac{1 \text{ mi}}{5280 \text{ ft}} = 3.25 \text{ mi}$

c. 3 in. $= \dfrac{3 \text{ in.}}{1} \cdot \dfrac{1 \text{ yd}}{36 \text{ in.}} = \dfrac{1}{12} \text{ yd}$

Your Turn (exercise #1, 5, 13, pg. 584):

Convert:

a. 30 in. to ft
b. 6 in. to yd
c. 23,760 ft to mi

188

- **Objective 3 - Convert units within the metric system.**

Solved Problem:

a. Convert 8000 meters to kilometers.
b. Convert 53 meters to millimeters.

a. 8000 m = 8 km
b. 53 m = 53,000 mm

Your Turn (exercise #17, 23, pg. 584):

a. Convert 5 m to cm.
b. Convert 0.023 mm to m.

- **Objective 4 - Use dimensional analysis to change to and from the metric system.**

Solved Problem:

TABLE 9.3 English and Metric Equivalents
1 inch (in.) = 2.54 centimeters (cm)
1 foot (ft) = 30.48 centimeters (cm)
1 yard (yd) ≈ 0.9 meter (m)
1 mile (mi) ≈ 1.6 kilometers (km)

Use **Table 9.3** (p. 582) to:

a. convert 8 feet to centimeters
b. convert 20 meters to yards
c. convert 30 meters to inches

a. $8 \text{ ft} = \dfrac{8 \text{ ft}}{1} \cdot \dfrac{30.48 \text{ cm}}{1 \text{ ft}} = 243.84 \text{ cm}$

b. $20 \text{ m} = \dfrac{20 \text{ m}}{1} \cdot \dfrac{1 \text{ yd}}{0.9 \text{ m}} \approx 22.22 \text{ yd}$

c. $30 \text{ m} = 3000 \text{ cm}$

$= \dfrac{3000 \text{ cm}}{1} \cdot \dfrac{1 \text{ in.}}{2.54 \text{ cm}}$

$\approx 1181.1 \text{ in.}$

Your Turn (exercise #27, 35, 43, pg. 584):

Use **Table 9.3** (p. 582) to

a. convert 14 in. to cm
b. convert 12 m to yd
c. convert 5 m to ft

Section 9.2 **Measuring Area and Volume**

- **Objective 1 - Use square units to measure area.**

Solved Problem:	**Your Turn (exercise #53, pg. 594):**
The population of California is 37,691,912 and its area is 158,633 square miles.	The population of Illinois is 12,869,257 and its area is 57, 914 square miles.
What is California's population density?	What is Illinois' population density?
Round to the nearest tenth.	Round to the nearest tenth.

$$= \frac{37,691,912 \text{ people}}{158,633 \text{ square miles}}$$

$$\approx 237.6 \text{ people per sq. mile}$$

- **Objective 2 - Use dimensional analysis to change units for area.**

Solved Problem:

The National Park Service administers approximately 84,000,000 acres of national parks.

How large is this in square miles?

84,000,000 acres

$$= \frac{84,000,000 \text{ acres}}{1} \cdot \frac{1 \text{ mi}^2}{640 \text{ acres}}$$

$$= \frac{84,000,000}{640} \text{ mi}^2 = 131,250 \text{ mi}^2$$

Your Turn (exercise #55, pg. 594):

The area of Everglades National Park (Florida) is 1,509,154 acres. How large is this in square miles?

192

- **Objective 3 - Use cubic units to measure volume.**

Solved Problem:

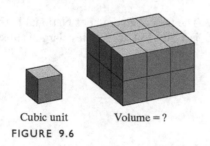

Cubic unit   Volume = ?

FIGURE 9.6

What is the volume of the region represented by the bottom row of blocks in **Figure 9.6** (p. 589)?

The volume is 9 cubic units.

Your Turn (exercise #13, pg. 593):

Find the volume of the figure in cubic units.

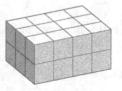

- **Objective 4 - Use English and metric units to measure capacity.**

Solved Problem:

TABLE 9.5 English Units for Capacity	
2 pints (pt) = 1 quart (qt)	
4 quarts = 1 gallon (gal)	
1 gallon = 128 fluid ounces (fl oz)	
1 cup (c) = 8 fluid ounces	
Volume in Cubic Units	**Capacity**
1 cubic yard	about 200 gallons
1 cubic foot	about 7.48 gallons
231 cubic inches	about 1 gallon

Use **Table 9.5** (p. 590).

A pool has a volume of 10,000 cubic feet. How many gallons of water does the pool hold?

$$10,000 \text{ ft}^3 = \frac{10,000 \text{ ft}^3}{1} \cdot \frac{7.48 \text{ gal}}{1 \text{ ft}^3} = 74,800 \text{ gal}$$

Your Turn (exercise #67, pg. 594):

A swimming pool has a volume of 45,000 cubic feet. How many gallons of water does the pool hold?

Section 9.3 **Measuring Weight and Temperature**

- **Objective 2 - Convert units of weight within the metric system.**

Solved Problem:

a. Convert 4.2 dg to mg.
b. Convert 620 cg to g.

a. 4.2 dg = 420 mg

b. 620 cg = 6.2 g

Your Turn (exercise #3, 5, pg. 602):

a. Convert 870 mg to g
b. Convert 8g to cg

- **Objective 3 - Use relationships between volume and weight within the metric system.**

Solved Problem:

Use **Table 9.10** (p. 598).

An aquarium holds 0.145 m³ of water. How much does the water weigh?

$$0.145 \text{ m}^3 = \frac{0.145 \, \text{m}^3}{1} \cdot \frac{1000 \text{ kg}}{1 \, \text{m}^3} = 145 \text{ kg}$$

Your Turn (exercise #11, pg. 602):

Use **Table 9.10** (p.598).

Convert 0.05cm³ to kg.

TABLE 9.10 Volume and Weight of Water in the Metric System					
Volume		**Capacity**		**Weight**	
1 cm³	=	1 mL	=	1 g	
1 dm³ = 1000 cm³	=	1 L	=	1 kg	
1 m³	=	1 kL	=	1000 kg = 1 t	

196

- **Objective 4 - Use dimensional analysis to change units of weight to and from the metric system.**

Solved Problem:

TABLE 9.11 Weight: English and Metric Equivalents
1 ounce (oz) ≈ 28 grams (g)
1 pound (lb) ≈ 0.45 kilogram (kg)
1 ton (T)   ≈ 0.9 tonne (t)

Using **Table 9.11** (p. 598),

a. Convert 120 pounds to kilograms.
b. Convert 500 grams to ounces.

a. $120 \text{ lb} \approx \dfrac{120 \text{ lb}}{1} \cdot \dfrac{0.45 \text{ kg}}{1 \text{ lb}} \approx 54 \text{ kg}$

b. $500 \text{ g} \approx \dfrac{500 \text{ g}}{1} \cdot \dfrac{1 \text{ oz}}{28 \text{ g}} \approx 17.9 \text{ oz}$

Your Turn (exercise #21, 23, pg. 603):

Use the following equivalents, along with dimensional analysis, to convert the given measurements to the unit indicated. When necessary, round answers to two decimal places.

$$16 \text{ oz} = 1 \text{ lb}$$
$$2000 \text{ lb} = 1 \text{ T}$$
$$1 \text{ oz} \approx 28 \text{g}$$
$$1 \text{ lb} \approx 0.45 \text{ kg}$$
$$1 \text{ T} \approx 0.9 \text{t}$$

a. 36 oz to g
b. 540 lb to kg

- **Objective 5 - Understand temperature scales.**

Solved Problem:	**Your Turn (exercise #33, 45, pg. 603):**
Convert $50°\,$C from $°$C to $°$F.	a. Convert $35°\,$C from $°$C to $°$F.
Convert $59°\,$F from $°$F to $°$C	
	b. Convert $23°\,$F from $°$F to $°$C.

$$F = \frac{9}{5} \cdot 50 + 32 = 122$$
$$50°C = 122°F$$

$$C = \frac{5}{9}(59 - 32) = 15$$
$$59°F = 15°C$$

198

Chapter 10
Geometry

Section 10.1 Points, Lines, Planes, and Angles

- **Objective 2 - Solve problems involving angle measures.**

Solved Problem:

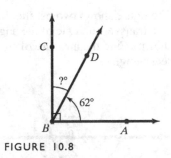

FIGURE 10.8

In **Figure 10.8** (p. 612) let $m\measuredangle DBC = 19°$. Find $m\measuredangle DBA$.

$90° - 19° = 71°$

Your Turn (exercise #11, pg. 616):

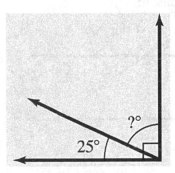

Find the measure of the angle marked with a question mark.

- **Objective 3 - Solve problems involving angles formed by parallel lines and transversals.**

Solved Problem:

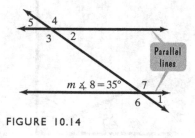

FIGURE 10.14

In **Figure 10.14** (p. 615), assume that $m\angle 8 = 29°$. Find the measure of each of the other seven angles.

$m\angle 1 = m\angle 8 = 29°$

$m\angle 5 = m\angle 8 = 29°$

$m\angle 2 = m\angle 8 = 29°$

$m\angle 6 = 180° - m\angle 8 = 180° - 29° = 151°$

$m\angle 7 = m\angle 6 = 151°$

$m\angle 3 = m\angle 7 = 151°$

$m\angle 4 = m\angle 3 = 151°$

Your Turn (exercise #29, pg. 616):

The figure shows two parallel lines intersected by a transversal. One of the angle measures is shown. Find the measure of each of the other seven angles.

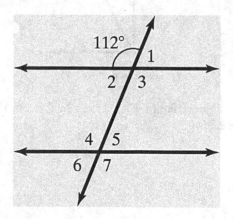

Section 10.2 Triangles

- **Objective 1 - Solve problems involving angle relationships in triangles.**

Solved Problem:

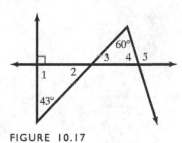

FIGURE 10.17

In **Figure 10.17** (p. 620), suppose that the angle shown to measure 43° measures, instead, 36°. Further suppose that the angle shown to measure 60° measures, instead, 58°. Under these new conditions, find the measures of angles 1 through 5 in the figure.

$m\angle 1 = 180° - 90° = 90°$

$m\angle 2 = 180° - 36° - m\angle 1$

$= 180° - 36° - 90°$

$= 54°$

$m\angle 3 = m\angle 2 = 54°$

$m\angle 4 = 180° - 58° - m\angle 3$

$= 180° - 58° - 54°$

$= 68°$

$m\angle 5 = 180° - m\angle 4$

$= 180° - 68°$

$= 112°$

Your Turn (exercise #5, pg. 626):

Find the measures of angles 1 through 5 in the figure shown.

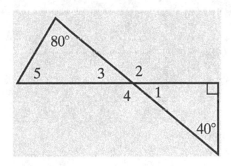

202

- **Objective 2 - Solve problems involving similar triangles.**

Solved Problem:

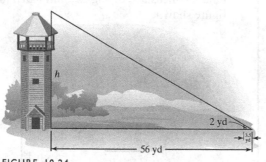

FIGURE 10.24

Find the height of the lookout tower shown in **Figure 10.24** (p. 623) using the figure that lines up the top of the tower with the top of a stick that is 2 yards long and 3.5 yards from the line to the top of the tower.

$$\frac{h}{2} = \frac{56}{3.5}$$
$$3.5 \cdot h = 2 \cdot 56$$
$$3.5h = 112$$
$$\frac{3.5h}{3.5} = \frac{112}{3.5}$$
$$h = 32 \text{ yd}$$

Your Turn (exercise #37, pg. 628):

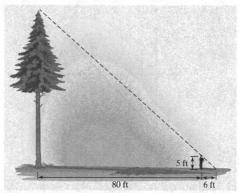

A person who is 5 feet tall is standing 80 feet from the base of a tree and the tree casts an 86-foot shadow. The person's shadow is 6 feet in length. What is the tree's height?

- **Objective 3 - Solve problems using the Pythagorean Theorem.**

Solved Problem:

Find the length of the hypotenuse in a right triangle whose legs have lengths 7 feet and 24 feet.

$$c^2 = a^2 + b^2$$
$$c^2 = 7^2 + 24^2$$
$$c^2 = 49 + 576$$
$$c^2 = 625$$
$$c = \sqrt{625}$$
$$c = 25 \text{ ft}$$

Your Turn (exercise #21, pg. 627):

Use the Pythagorean Theorem to find the missing length in the right triangle. Use your calculator to find square roots, rounding, if necessary, to the nearest tenth.

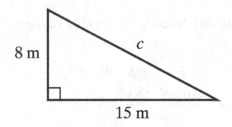

204

Section 10.3 Polygons, Perimeter, and Tessellations

- **Objective 3 - Solve problems involving a polygon's perimeter.**

Solved Problem:

A rectangular field has a length of 50 yards and a width of 30 yards. If fencing costs $6.50 per foot, find the cost to enclose the field with fencing.

Note: 50 yds equals 150 ft and 30 yds equals 90 ft

$P = 2l + 2w$

$P = 2 \cdot 150 \text{ ft} + 2 \cdot 90 \text{ ft} = 480 \text{ ft}$

$\text{Cost} = \dfrac{480 \text{ feet}}{1} \cdot \dfrac{\$6.50}{\text{foot}} = \$3120$

Your Turn (exercise #47, pg. 638):

A school playground is in the shape of a rectangle 400 feet long and 200 feet wide. If fencing costs $14 per yard, what will it cost to place fencing around the playground?

- **Objective 4 - Find the sum of the measures of a polygon's angles.**

Solved Problem:

Find the sum of the measures of the angles of a 12-sided polygon.

Sum $= (n-2)180°$
$= (12-2)180°$
$= 10 \cdot 180°$
$= 1800°$

Your Turn (exercise #25, pg. 637):

Find the sum of the measures of the angles of a five-sided polygon.

- **Objective 5 - Understand tessellations and their angle requirements.**

Solved Problem:

Explain why a tessellation cannot be created using only regular octagons.

Each angle is $\dfrac{(n-2)\,180°}{n} = \dfrac{(8-2)\,180°}{8} = 135°$

Regular octagons cannot be used to create a tessellation because $360°$ is not a multiple of $135°$.

Your Turn (exercise #37, pg. 637):

Can a tessellation be created using only regular nine-sided polygons? Explain your answer.

- **Objective 1 - Use area formulas to compute the areas of plane regions and solve applied problems.**

Solved Problem:

What will it cost to carpet a rectangular floor measuring 18 feet by 21 feet if the carpet costs $16 per square yard?

First convert the linear measures in feet to linear yards.

$$18 \text{ ft} = \frac{18 \text{ ft}}{1} \cdot \frac{1 \text{ yd}}{3 \text{ ft}} = 6 \text{ yd}$$

$$21 \text{ ft} = \frac{21 \text{ ft}}{1} \cdot \frac{1 \text{ yd}}{3 \text{ ft}} = 7 \text{ yd}$$

Area of floor $= 6 \text{ yd} \cdot 7 \text{ yd} = 42 \text{ yd}^2$

$$\text{Cost of carpet} = \frac{42 \text{ yd}^2}{1} \cdot \frac{\$16}{1 \text{ yd}^2} = \$672$$

Your Turn (exercise #37, pg. 648):

What will it cost to carpet a rectangular floor measuring 9 feet by 21 feet if the carpet costs $26.50 per square yard?

208

- **Objective 2 - Use formulas for a circle's circumference and area.**

Solved Problem:

Find the circumference of a circle whose diameter measures 10 inches. Express the answer in terms of π and then round to the nearest tenth of an inch.

$C = \pi d$
$\quad = \pi(10 \text{ in.})$
$\quad = 10\pi \text{ in.}$

$\pi \approx 3.14$
$C \approx 3.14d$
$\quad \approx 10(3.14) \text{ in.}$
$\quad \approx 31.4 \text{ in.}$

Your Turn (exercise #17, pg. 647):

Find the circumference and area of the circle. Express the answer in terms of π and then round to the nearest tenth of an inch.

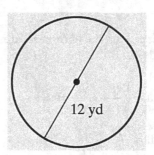

12 yd

Section 10.5 Volume and Surface Area

- **Objective 1 - Use volume formulas to compute the volumes of three-dimensional figures and solve applied problems.**

Solved Problem:

Find the volume of a rectangular solid with length 5 feet, width 3 feet, and height 7 feet.

$V = 5 \text{ ft} \cdot 3 \text{ ft} \cdot 7 \text{ ft} = 105 \text{ ft}^3$

Your Turn (exercise #1, pg. 656):

Find the volume of the figure.

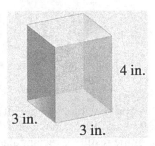

210

- **Objective 2 - Compute the surface area of a three-dimensional figure.**

Solved Problem:

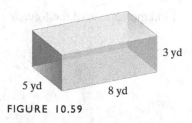

FIGURE 10.59

If the length, width, and height shown in **Figure 10.59** (p. 655) are each doubled, find the surface area of the resulting rectangular solid.

New dimensions: $l = 16$ yd, $w = 10$ yd, $h = 6$ yd
$$SA = 2lw + 2lh + 2wh$$
$$= 2 \cdot 16 \text{ yd} \cdot 10 \text{ yd} + 2 \cdot 16 \text{ yd} \cdot 6 \text{ yd} + 2 \cdot 10 \text{ yd} \cdot 6 \text{ yd}$$
$$= 320 \text{ yd}^2 + 192 \text{ yd}^2 + 120 \text{ yd}^2$$
$$= 632 \text{ yd}^2$$

Your Turn (exercise #21, pg. 657):

Find the surface area of the figure.

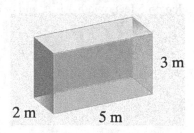

Section 10.6 Right Triangle Trigonometry

- **Objective 1 - Use the lengths of the sides of a right triangle to find trigonometric ratios.**

Solved Problem:	**Your Turn (exercise #3, pg. 665):**
Find the sine, cosine, and tangent of A in the figure shown.	Use the given right triangle to find ratios, in reduced form, for $\sin A$, $\cos A$, and $\tan A$.

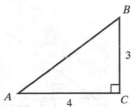

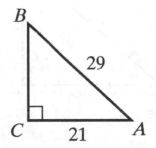

Begin by finding the measure of the hypotenuse c using the Pythagorean Theorem.

$$c^2 = a^2 + b^2 = 3^2 + 4^2 = 25$$
$$c = \sqrt{25} = 5$$

$$\sin A = \frac{3}{5}$$

$$\cos A = \frac{4}{5}$$

$$\tan A = \frac{3}{4}$$

212

- **Objective 2 - Use trigonometric ratios to find missing parts of right triangles.**

Solved Problem:

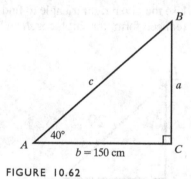

FIGURE 10.62

In **Figure 10.62** (p. 661), let $m\angle A = 62°$ and $b = 140$ cm. Find a to the nearest centimeter.

$$\tan A = \frac{a}{b}$$

$$\tan 62° = \frac{a}{140}$$
$$a = 140 \tan 62° \approx 263 \text{ cm}$$

Your Turn (exercise #9, pg. 665):

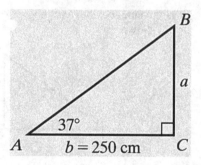

Find the measure of the side of the right triangle whose length is designated by a lowercase letter. Round answer to the nearest whole number.

- **Objective 3 - Use trigonometric ratios to solve applied problems.**

Solved Problem:

A flagpole that is 14 meters tall casts a shadow 10 meters long. Find the angle of elevation of the sun to the nearest degree.

$$\tan A = \frac{14}{10}$$

$$A = \tan^{-1}\left(\frac{14}{10}\right) \approx 54°$$

Your Turn (exercise #47, pg. 668):

A wheelchair ramp is to be built beside the steps to the campus library. Find the angle of elevation of the 23-foot ramp, to the nearest tenth of a degree, if its final height is 6 feet.

Section 10.7 Beyond Euclidean Geometry

- **Objective 1 - Gain an understanding of some of the general ideas of other kinds of geometries.**

Solved Problem:

Create a graph with two even and two odd vertices. Then describe a path that will traverse it.

Answers will vary. Possible answer:

The upper left and lower right vertices are odd. The lower left and upper right vertices are even.

One possible tracing:
Start at the upper left, trace around the square, then trace down the diagonal.

Your Turn (exercise #1, pg. 676):

For the graph,

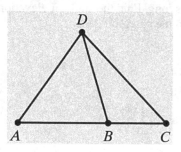

a. Is the graph transversable?

b. If it is, describe a path that will transverse it.

Chapter 11
Counting Methods and Probability Theory

Section 11.1 The Fundamental Counting Principle

- **Objective 1 - Use the Fundamental Counting Principle to determine the number of possible outcomes in a given situation.**

Solved Problem:

A restaurant offers 10 appetizers and 15 main courses. In how many ways can you order a two course meal?

Multiply the number of choices for eac

h of the two courses of the meal:

Appetizers: Main Courses:

   $10 \quad \times \quad 15 \quad = 150$

Your Turn (exercise #1, pg. 692):

A restaurant offers eight appetizers and ten main courses. In how many ways can a person order a two-course meal?

216

- **Objective 1 - Use the Fundamental Counting Principle to count permutations.**

Solved Problem:

In how many ways can you arrange five books along a shelf, assuming that the order of the books makes a difference?

The number of choices decreases by 1 each time a book is selected.

1st Book: 2nd: 3rd: 4th: 5th:

    5    $\times$ 4 $\times$ 3 $\times$ 2 $\times$ 1 $= 120$

Your Turn (exercise #1, pg. 700):

Six performers are to present their comedy acts on a weekend evening at a comedy club. How many different ways are there to schedule their appearances?

- **Objective 2 - Evaluate factorial expressions.**

Solved Problem:

Evaluate without using a calculator's factorial key [from Check Point 3, p. 697]:

a. $\dfrac{9!}{6!}$ c. $\dfrac{100!}{99!}$

a. $\dfrac{9!}{6!} = \dfrac{9 \cdot 8 \cdot 7 \cdot 6!}{6!} = \dfrac{9 \cdot 8 \cdot 7 \cdot \cancel{6!}}{\cancel{6!}} = 9 \cdot 8 \cdot 7 = 504$

c. $\dfrac{100!}{99!} = \dfrac{100 \cdot 99!}{99!} = \dfrac{100 \cdot \cancel{99!}}{\cancel{99!}} = 100$

Your Turn (exercise #15, 21, pg. 701):

Evaluate each factorial expression.

$\dfrac{29!}{25!}$

$\dfrac{104!}{102!}$

218

- **Objective 3 - Use the permutations formula.**

Solved Problem:

A corporation has seven members on its board of directors. In how many different ways can it elect a president, vice-president, secretary, and treasurer?

$$_7P_4 = \frac{7!}{(7-4)!} = \frac{7!}{3!}$$

$$= \frac{7 \cdot 6 \cdot 5 \cdot 4 \cdot 3!}{3!}$$

$$= \frac{7 \cdot 6 \cdot 5 \cdot 4 \cdot \cancel{3!}}{\cancel{3!}}$$

$$= 7 \cdot 6 \cdot 5 \cdot 4$$

$$= 840$$

Your Turn (exercise #41, pg. 701):

A club with ten members is to choose three officers – president, vice-president, and secretary-treasurer. If each office is to be held by one person and no person can hold more than one office, in how many ways can those offices be filled?

- **Objective 4 - Find the number of permutations of duplicate items**.

Solved Problem:

In how many ways can the letters of the word **OSMOSIS** be arranged?

There are 7 letters with 2 O's and 3 S's. Thus,

$$\frac{n!}{p!q!} = \frac{7!}{2!3!} = \frac{7\cdot6\cdot5\cdot4\cdot\cancel{3!}}{2\cdot1\cdot\cancel{3!}} = 420$$

Your Turn (exercise #51, pg. 701):

How many distinct permutations can be formed using the letters of the word **TALLAHASSEE**?

Section 11.3 Combinations

- **Objective 1 - Distinguish between permutation and combination problems.**

Solved Problem:

For each of the following problems, determine whether the problem is one involving permutations or combinations. (It is not necessary to solve the problem.)

a. How many ways can you select 6 free DVDs from a list of 200 DVDs?

b. In a race in which there are 50 runners and no ties, in how many ways can the first three finishers come in?

a. The order in which you select the DVDs does not matter. This problem involves combinations.

b. Order matters. This problem involves permutations.

Your Turn (exercise #3, pg. 707):

For the following problem, determine whether the problem is one involving permutations or combinations. (It is not necessary to solve the problem).

How many different four-letter passwords can be formed from the letters A, B, C, D, E, F, and G if no repetition of letters is allowed?

- **Objective 2 - Solve problems involving combinations using the combinations formula.**

Solved Problem:

You volunteer to pet-sit for your friend who has seven different animals. How many different pet combinations are possible if you take three of the seven pets?

$$_7C_3 = \frac{7!}{(7-3)!3!} = \frac{7!}{4!3!}$$
$$= \frac{7\cdot6\cdot5\cdot4!}{4!\cdot3\cdot2\cdot1}$$
$$= \frac{7\cdot6\cdot5\cdot \cancel{4!}}{\cancel{4!}\cdot3\cdot2\cdot1}$$
$$= \frac{7\cdot6\cdot5}{3\cdot2\cdot1}$$
$$= 35$$

Your Turn (exercise #29, pg. 707):

An election ballot asks voters to select three city commissioners from a group of six candidates. In how many ways can this be done?

Section 11.4 Fundamentals of Probability

- **Objective 1 - Compute theoretical probability.**

Solved Problem:	**Your Turn (exercise #1, 3, 7, 9, pg. 715):**
A six-sided die is rolled once. Find the probability of rolling	A six-sided die is rolled once. Find the probability of rolling
a. a 2	a. a 4
b. a number less than 4	b. an odd number
c. a number greater than 7	c. a number less than 20
d. a number less than 7	d. a number greater than 20

$S = \{1, 2, 3, 4, 5, 6\}$

a. The event of getting a 2 can occur in one way.

$$P(2) = \frac{\text{number of ways a 2 can occur}}{\text{total number of possible outcomes}} = \frac{1}{6}$$

b. The event of getting a number less than 4 can occur in three ways: 1, 2, 3.

$P(\text{less than 4})$

$$= \frac{\text{number of ways a number less than 4 can occur}}{\text{total number of possible outcomes}}$$

$$= \frac{3}{6} = \frac{1}{2}$$

c. The event of getting a number greater than 7 cannot occur.

$P(\text{greater than 7})$

$$= \frac{\text{number of ways a number greater than 7 can occur}}{\text{total number of possible outcomes}}$$

$$= \frac{0}{6} = 0$$

The probability of an event that cannot occur is 0.

d. The event of getting a number less than 7 can occur in six ways: 1, 2, 3, 4, 5, 6.

$P(\text{less than 7})$

$$= \frac{\text{number of ways a number less than 7 can occur}}{\text{total number of possible outcomes}}$$

$$= \frac{6}{6} = 1$$

The probability of any certain event is 1.

- **Objective 2 - Compute empirical probability.**

Solved Problem:

If one person is randomly selected from the population described in **Table 11.3** (p. 714), find the probability, expressed as a decimal rounded to the nearest hundredth, that the person

a. has never been married.
b. is male.

a. P(never married)

$= \dfrac{\text{number of persons never married}}{\text{total number of U.S. adults}} = \dfrac{74}{242} \approx 0.31$

b. P(male)

$= \dfrac{\text{number of males}}{\text{total number of U.S. adults}} = \dfrac{118}{242} \approx 0.49$

Your Turn (exercise #61, 63, pg. 717):

Use the table to find the probability, expressed as a decimal rounded to the nearest hundredth, that a randomly selected American who moved was

a. an owner
b. a person who moved within the same state

NUMBER OF PEOPLE IN THE UNITED STATES WHO MOVED, IN MILLIONS

	Moved to Same State	Moved to Different State	Moved to Different Country
Owner	11.7	2.8	0.3
Renter	18.7	4.5	1.0

TABLE 11.3 Marital Status of the U.S. Population, Ages 15 or Older, 2010, in Millions

	Married	Never Married	Divorced	Widowed	Total
Male	65	40	10	3	118
Female	65	34	14	11	124
Total	130	74	24	14	242

Total male: $65 + 40 + 10 + 3 = 118$

Total female: $65 + 34 + 14 + 11 = 124$

Total married: $65 + 65 = 130$

Total never married: $40 + 34 = 74$

Total divorced: $10 + 14 = 24$

Total widowed: $3 + 11 = 14$

Total adult population: $118 + 124 = 242$

Source: U.S. Census Bureau

Section 11.5 Probability with the Fundamental Counting Principle, Permutations, and Combinations

- **Objective 1 - Compute probabilities with permutations.**

Solved Problem:

Consider the six jokes about books by Groucho Marx, Bob Blitzer, Steven Wright, Henny Youngman, Jerry Seinfeld, and Phyllis Diller. As in Example 1 (p. 718), each joke is written on one of six cards which are randomly drawn one card at a time. The order in which the cards are drawn determines the order in which the jokes are delivered.

What is the probability that a joke by a comic whose first name begins with G is told first and a man's joke is delivered last?

The total number of permutations = 6!
$= 6 \cdot 5 \cdot 4 \cdot 3 \cdot 2 \cdot 1 = 720$

For the given outcome there is 1 choice (first name beginning with G) for the first joke, which would leave 4 choices for the last joke (the 4 remaining men). The remaining jokes have 4, 3, 2, and 1 choice respectively.

Your Turn (exercise #1, pg. 723):

Martha, Lee, Nancy, Paul, and Armando have all been invited to a dinner party. They arrive randomly, and each person arrives at a different time.

a. In how many ways can they arrive?
b. In how many ways can Martha arrive first and Armando last?
c. Find the probability that Martha will arrive first and Armando last.

First name begins with G   4 jokes other than first and last   man (other than first joke)

1st:		2nd:	3rd:		4th:		5th:		6th:	
1	×	4	× 3	×	2	×	1	×	4	= 96

$P(\text{first joke is by a man whose name begins with G and the last is by a man}) = \dfrac{96}{720} = \dfrac{2}{15}$

- **Objective 2 - Compute probabilities with combinations.**

Solved Problem:

A club consists of six men and four women. Three members are selected at random to attend a conference. Find the probability that the selected group consists of two men and one woman.

The total number of combinations:

$$_{10}C_3 = \frac{10!}{(10-3)!3!} = \frac{10!}{7!3!} = \frac{10 \cdot 9 \cdot 8 \cdot 7!}{7!3 \cdot 2 \cdot 1} = 120$$

Select 2 out of 6 men:

$$_6C_2 = \frac{6!}{(6-2)!2!} = \frac{6!}{4!2!} = \frac{6 \cdot 5 \cdot 4!}{4!2 \cdot 1} = 15$$

Select 1 out of 4 women:

$$_4C_1 = \frac{4!}{(4-1)!1!} = \frac{4!}{3!1!} = \frac{4 \cdot 3!}{3!} = \frac{4 \cdot 3!}{3!} = 4$$

total number of combinations of 2 men and 1 woman:
$15 \times 4 = 60$

$P(2 \text{ men}, 1 \text{ woman})$

$$= \frac{\text{number of combinations with 2 men, 1 woman}}{\text{total number of combinations}}$$

$$= \frac{60}{120} = \frac{1}{2}$$

Your Turn (exercise #13, pg. 724):

A city council consists of six Democrats and four Republicans. If a committee of three people is selected, find the probability of selecting one Democrat and two Republicans.

228

- **Objective 1 - Find the probability that an event will not occur.**

Solved Problem:

If you are dealt one card from a standard 52-card deck, find the probability that you are not dealt a diamond.

$P(\text{not a diamond}) = 1 - P(\text{diamond})$

$$= 1 - \frac{13}{52}$$

$$= \frac{39}{52} = \frac{3}{4}$$

Your Turn (exercise #1, pg. 734):

If you are dealt one card from a standard 52-card deck, find the probability that you are not dealt an ace.

- **Objective 2 - Find the probability of one event or a second event occurring.**

Solved Problem:

If you roll a single, six-sided die, what is the probability of getting either a 4 or a 5?

$$P(4 \text{ or } 5) = P(4) + P(5)$$
$$= \frac{1}{6} + \frac{1}{6}$$
$$= \frac{2}{6} = \frac{1}{3}$$

Your Turn (exercise #19, pg. 735):

If you randomly select one card from a standard 52-card deck, find the probability of selecting a red 2 or a black 3.

230

- **Objective 3 - Understand and use odds.**

Solved Problem:

You are dealt one card from a 52-card deck.

a. Find the odds in favor of getting a red queen.
b. Find the odds against getting a red queen.

There are 2 red queens. Number of favorable outcomes = 2, Number of unfavorable outcomes = 50

a. Odds in favor of getting a red queen are 2 to 50 or 2:50 which reduces to 1:25.

b. Odds against getting a red queen are 50 to 2 or 50:2 which reduces to 25:1.

Your Turn (exercise #61, 63, pg. 736):

A single die is rolled. Find the odds

a. in favor of rolling a number greater than 2
b. against rolling a number greater than 2

Section 11.7 Events Involving And; Conditional Probability

- **Objective 1 - Find the probability of one event and a second event occurring.**

Solved Problem:

If the probability that South Florida will be hit by a hurricane in any single year is $\frac{5}{19}$,

a. What is the probability that South Florida will be hit by a hurricane in four consecutive years?
b. What is the probability that South Florida will not be hit by a hurricane in the next four years?
c. What is the probability that South Florida will be hit by a hurricane at least once in the next four years?

Express all probabilities as fractions and as decimals rounded to three places.

a. P(hit four years in a row)

$= P(\text{hit}) \cdot P(\text{hit}) \cdot P(\text{hit}) \cdot P(\text{hit})$

$= \dfrac{5}{19} \cdot \dfrac{5}{19} \cdot \dfrac{5}{19} \cdot \dfrac{5}{19}$

$= \dfrac{625}{130,321} \approx 0.005$

b. Note: P(not hit in any single year)

$= 1 - P(\text{hit in any single year})$

$= 1 - \dfrac{5}{19}$

$= \dfrac{14}{19}$

Therefore,
P(not hit in next four years)

$= P(\text{not hit}) \cdot P(\text{not hit}) \cdot P(\text{not hit}) \cdot P(\text{not hit})$

$= \dfrac{14}{19} \cdot \dfrac{14}{19} \cdot \dfrac{14}{19} \cdot \dfrac{14}{19}$

$= \dfrac{38,416}{130,321} \approx 0.295$

c. P(hit at least once in next four years)

$= 1 - P(\text{not hit in next four years})$

$= 1 - \dfrac{38,416}{130,321}$

$= \dfrac{91,905}{130,321} \approx 0.705$

Your Turn (exercise #25b, c, d, pg. 746):

The probability that South Florida will be hit by a major hurricane (category 4 or 5) in any single year is $\frac{1}{16}$.

a. What is the probability that South Florida will be hit by a major hurricane in three consecutive years?
b. What is the probability that South Florida will not be hit by a major hurricane in the next ten years?
c. What is the probability that South Florida will be hit by a major hurricane at least once in the next ten years?

- **Objective 2 - Compute conditional probabilities.**

Solved Problem:

Use the data in **Table 11.7** (p. 744) to find the probability that a U.S. woman aged 40 to 49

a. has a positive mammogram, given that she has breast cancer.
b. has breast cancer, given that she has a positive mammogram.

Express probabilities as decimals and, if necessary, round to three decimal places.

a. $P(\text{positive mammogram}|\text{breast cancer})$

$=\dfrac{720}{800}$

$=\dfrac{9}{10}=0.9$

b. $P(\text{breast cancer}|\text{positive mammogram})$

$=\dfrac{720}{7664}$

$=\dfrac{45}{479}=0.094$

Your Turn (exercise #57, 59, pg. 747):

Use the data in the table to

a. find the probability of surviving a car accident, given that the driver wore a seat belt.

b. find the probability of wearing a seat belt, given that the driver survived a car accident.

CAR ACCIDENTS IN FLORIDA

	Wore Seat Belt	No Seat Belt	Total
Driver Survived	412,368	162,527	574,895
Driver Died	510	1601	2111
Total	412,878	164,128	577,006

TABLE 11.7 Mammography Screening on 100,000 U.S. Women, Ages 40 to 49			
	Breast Cancer	**No Breast Cancer**	**Total**
Positive Mammogram	720	6944	7664
Negative Mammogram	80	92,256	92,336
Total	800	99,200	100,000

Source: Gerd Gigerenzer, *Calculated Risks*. Simon and Schuster, 2002.

Section 11.8 Expected Value

- **Objective 1 - Compute expected value.**

Solved Problem:

It is equally probable that a pointer will land on any one of four regions, numbered 1 through 4. Find the expected value for where the pointer will stop.

$$E = 1 \cdot \frac{1}{4} + 2 \cdot \frac{1}{4} + 3 \cdot \frac{1}{4} + 4 \cdot \frac{1}{4}$$

$$= \frac{1+2+3+4}{4}$$

$$= \frac{10}{4} = 2.5$$

Your Turn (exercise #1, pg. 754):

The numbers that the pointer can land on and their respective probabilities are shown. Compute the expected value for the number on which the pointer lands.

Outcome	Probability
1	$\frac{1}{2}$
2	$\frac{1}{4}$
3	$\frac{1}{4}$

1.75

234

• **Objective 2 - Use expected value to solve applied problems.**

Solved Problem:

Work Example 3 again (p. 750) if the probabilities for claims of $0 and $10,000 are reversed. Thus, the probability of a $0 claim is 0.01 and the probability of a $10,000 claim is 0.70.

TABLE 11.10 Probabilities for Auto Claims	
Amount of Claim (to the nearest $2000)	**Probability**
$0	0.70
$2000	0.15
$4000	0.08
$6000	0.05
$8000	0.01
$10,000	0.01

Example 3 reads: An automobile insurance company has determined the probabilities for various claim amounts for drivers ages 16 through 21, shown in **Table 11.10** (p. 750).

a. Calculate the expected value and describe what this means in practical terms.

b. How much should the company charge as an average premium so that it does not lose or gain money on its claim costs?

a. $E = \$0(0.01) + \$2000(0.15) + \$4000(0.08) + \$6000(0.05) + \$8000(0.01) + \$10,000(0.70) = \$8000$
This means that in the long run, the average cost of a claim is expected to be $8000.

b. An average premium charge of $8000 would cause the company to neither lose nor gain money.

Your Turn (exercise #3a, b, pg. 754):

The table shows claims and their probabilities for an insurance company.

PROBABILITIES FOR HOMEOWNERS' INSURANCE CLAIMS

Amount of Claim (to the nearest $50,000)	**Probability**
$0	0.65
$50,000	0.20
$100,000	0.10
$150,000	0.03
$200,000	0.01
$250,000	0.01

a. Calculate the expected value and describe what this means in practical terms.

b. How much should the company charge as an average premium so that it breaks even on its claim costs?

- **Objective 3 - Use expected value to determine the average payoff or loss in a game of chance.**

Solved Problem:

A charity is holding a raffle and sells 1000 raffle tickets for $2 each. One of the tickets will be selected to win a grand prize of $1000. Two other tickets will be selected to win consolation prizes of $50 each. Fill in the missing column in **Table 11.14** (p. 753). Then find the expected value if you buy one raffle ticket. Describe what this means in practical terms. What can you expect to happen if you purchase five tickets?

Values of gain or loss:
Grand Prize: $1000 - $2 = 998,
Consolation Prize: $50 - $2 = 48,
Nothing: $0 - $2 = -$2$

$$E = (-\$2)\left(\frac{997}{1000}\right) + (\$48)\left(\frac{2}{1000}\right) + (\$998)\left(\frac{1}{1000}\right)$$

$$= \frac{-\$1994 + \$96 + \$998}{1000}$$

$$= -\frac{\$900}{1000} = -\$0.90$$

The expected value for one ticket is $-\$0.90$. This means that in the long run a player can expect to lose $0.90 for each ticket bought. Buying five tickets will make your likelihood of winning five times greater, however there is no advantage to this strategy because the *cost* of five tickets is also five times greater than one ticket.

Your Turn (exercise #15, pg. 755):

A game is played using one die. If the die is rolled and shows 1, the player wins $5. If the die shows any number other than 1, the player wins nothing. If there is a charge of $1 to play the game, what is the game's expected value? What does this value mean?

TABLE 11.14 Gains, Losses, and Probabilities in a Raffle		
Outcome	**Gain or Loss**	**Probability**
Win Grand Prize		$\frac{1}{1000}$
Win Consolation Prize		$\frac{2}{1000}$
Win Nothing		$\frac{997}{1000}$

236

Chapter 12
Statistics

Section 12.1 Sampling, Frequency Distributions, and Graphs

- **Objective 2 - Select an appropriate sampling technique.**

Solved Problem:

A city government wants to conduct a survey among the city's homeless to discover their opinions about required residence in city shelters from midnight until 6 a.m.

Explain why the sampling technique described [from Check Point 1b on page 767] is not a random sample. Then describe an appropriate way to select a random sample of the city's homeless.

The sampling technique described in Check Point 1b does not produce a random sample because homeless people who do not go to shelters have no chance of being selected for the survey. In this instance, an appropriate method would be to randomly select neighborhoods of the city and then randomly survey homeless people within the selected neighborhood.

Your Turn (exercise #1, pg. 776):

The government of a large city needs to determine whether the city's residents will support the construction of a new jail. The government decides to conduct a survey of a sample of the city's residents. Which one of the following procedures would be most appropriate for obtaining a sample of the city's residents?

a. Survey a random sample of the employees and inmates at the old jail.

b. Survey every fifth person who walks into City Hall on a given day.

c. Survey a random sample of persons within each geographic region of the city.

d. Survey the first 200 people listed in the city's telephone directory.

238

- **Objective 3 - Organize and present data.**

Solved Problem:

Construct a frequency distribution for the data showing final course grades for students in a precalculus course, listed alphabetically by student name in a grade book:

F, A, B, B, C, C, B, C, A, A,
C, C, D, C, B, D, C, C, B, C.

Grade	Number of students
A	3
B	5
C	9
D	2
F	1
	20

Your Turn (exercise #7, pg. 776):

A random sample of 30 college students is selected. Each student is asked how much time her or she spent on homework during the previous week. The following times (in hours) are obtained:

16, 24, 18, 21, 18, 16, 18, 17, 15, 21, 19, 17, 17, 16, 19, 18, 15, 15, 20, 17, 15, 17, 24, 19, 16, 20, 16, 19, 18, 17

Construct a frequency distribution for the data.

Section 12.2 Measures of Central Tendency

- **Objective 1 - Determine the mean for a data set.**

Solved Problem:

Use **Table 12.6** (p. 781) to find the mean earnings, x, in millions of dollars, for the ten highest-earning actresses.

$$\text{Mean} = \frac{\sum x}{n}$$

$$= \frac{13+13+10+10+10+9+9+7+7+7}{10}$$

$$= \frac{95}{10} = 9.5$$

$9.5 million

Your Turn (exercise #7, pg. 791):

Find the mean for the following group of data items:

1.6, 3.8, 5.0, 2.7, 4.2, 4.2, 3.2, 4.7, 3.6, 2.5, 2.5

TABLE 12.6 Highest-Earning TV Actors and Actresses, 2010–2011			
Actor	**Earnings (millions of dollars)**	**Actress**	**Earnings (millions of dollars)**
Charlie Sheen	$40	Eva Longoria	$13
Ray Romano	$20	Tina Fey	$13
Steve Carell	$15	Marcia Cross	$10
Mark Harmon	$13	Mariska Hargitay	$10
Jon Cryer	$11	Marg Helgenberger	$10
Laurence Fishburne	$11	Teri Hatcher	$9
Patrick Dempsey	$10	Felicity Huffman	$9
Simon Baker	$9	Courteney Cox	$7
Hugh Laurie	$9	Ellen Pompeo	$7
Chris Meloni	$9	Julianna Margulies	$7

Source: Forbes

240

- **Objective 2 - Determine the median for a data set.**

Solved Problem:

Find the median for each of the following groups of data:

a. 28, 42, 40, 25, 35

b. 72, 61, 85, 93, 79, 87

a. First arrange the data items from smallest to largest: 25, 28, <u>35</u>, 40, 42

The number of data items is odd, so the median is the middle number. The median is 35.

b. First arrange the data items from smallest to largest: 61, 72, <u>79</u>, <u>85</u>, 87, 93

The number of data items is even, so the median is the mean of the two middle data items.

The median is $\dfrac{79+85}{2} = \dfrac{164}{2} = 82$.

Your Turn (exercise #15, 19, pg. 791):

Find the median for each of the following groups of data:

a. 91, 95, 99, 97, 93, 95

b. 1.6, 3.8, 5.0, 2.7, 4.2, 4.2, 3.2, 4.7, 3.6, 2.5, 2.5

- **Objective 3 - Determine the mode for a data set.**

Solved Problem:

Find the mode for each of the following groups of data:

a. 3, 8, 5, 8, 9, 10

b. 3, 8, 5, 8, 9, 3

c. 3, 8, 5, 6, 9, 10

a. The mode is 8 (because 8 occurs most often).

b. The modes are 3 and 8 (because both 3 and 8 occur most often).

c. There is no mode (because each data item occurs the same number of times).

Your Turn (exercise #25, 29, 31, pg. 791):

Find the mode for each of the following groups of data:

a. 7, 4, 3, 2, 8, 5, 1, 3

b. 100, 40, 70, 40, 60

c. 1.6, 3.8, 5.0, 2.7, 4.2, 4.2, 3.2, 4.7, 3.6, 2.5, 2.5

- **Objective 4 - Determine the midrange for a data set.**

Solved Problem:

Use **Table 12.12** on the previous page to find the midrange score among the 12 worst countries to be a woman.

$$\text{Midrange} = \frac{0.0 + 29.0}{2} = 14.5$$

Your Turn (exercise #43, pg. 791):

Find the midrange for the following group of data items:

1.6, 3.8, 5.0, 2.7, 4.2, 4.2, 3.2, 4.7, 3.6, 2.5, 2.5

TABLE 12.12 Women in the World			
Best Places to Be a Woman		**Worst Places to Be a Woman**	
Country	**Score**	**Country**	**Score**
Iceland	100.0	Chad	0.0
Canada	99.6	Afghanistan	2.0
Sweden	99.2	Yemen	12.1
Denmark	95.3	Democratic Republic of the Congo	13.6
Finland	92.8	Mali	17.6
Switzerland	91.9	Solomon Islands	20.8
Norway	91.3	Niger	21.2
United States	89.8	Pakistan	21.4
Australia	88.2	Ethiopia	23.7
Netherlands	87.7	Sudan	26.1
New Zealand	87.2	Guinea	28.5
France	87.2	Sierra Leone	29.0

Source: Newsweek

Section 12.3 Measures of Dispersion

- **Objective 1 - Determine the range for a data set.**

Solved Problem:

Find the range for the following group of data items:

4, 2, 11, 7

$$\text{Range} = 11 - 2 = 9$$

Your Turn (exercise #5, pg. 800):

Find the range for the following group of data items:

3, 3, 4, 4, 5, 5

244

- **Objective 2 - Determine the standard deviation for a data set.**

Solved Problem:

Find the standard deviation for the following group of data items from Check Point 2 on page 796:

2, 4, 7, 11.

Round to two decimal places.

$$\text{Mean} = \frac{2+4+7+11}{4} = \frac{24}{4} = 6$$

Data item	Deviation: Data item – mean	$(\text{Deviation})^2$: $(\text{Data item–mean})^2$
2	$2 - 6 = -4$	$(-4)^2 = 16$
4	$4 - 6 = -2$	$(-2)^2 = 4$
7	$7 - 6 = 1$	$1^2 = 1$
11	$11 - 6 = 5$	$5^2 = 25$

$$\sum (\text{data item–mean})^2 = 46$$

$$\text{Standard deviation} = \sqrt{\frac{46}{4-1}} = \sqrt{\frac{46}{3}} \approx 3.92$$

Your Turn (exercise #17, pg. 800):

Find the standard deviation for the following group of data items. Round answer to two decimal places.

1, 2, 3, 4, 5

Section 12.4 The Normal Distribution

- **Objective 3 - Find scores at a specified standard deviation from the mean.**

Solved Problem:

Female adult heights in North America are approximately normally distributed with a mean of 65 inches and a standard deviation of 3.5 inches. Find the height that is

a. 3 standard deviations above the mean.
b. 2 standard deviations below the mean.

a. Height = mean + 3 · standard deviation
$$= 65 + 3 \cdot 3.5 = 75.5 \text{ in.}$$

b. Height = mean − 2 · standard deviation
$$= 65 - 2 \cdot 3.5 = 58 \text{ in.}$$

Your Turn (exercise #2, 8, pg. 813):

The scores on a test are normally distributed with a mean of 100 and a standard deviation of 20. Find the score that is

a. 2 standard deviations above the mean

b. 3 standard deviations below the mean

246

- **Objective 4 - Use the 68–95–99.7 Rule.**

Solved Problem:

Use the distribution of male adult heights in North America in **Figure 12.12** (p. 804) to find the percentage of men with heights

a. between 62 inches and 78 inches.
b. between 70 inches and 78 inches.
c. above 74 inches.

a. The 68-95-99.7 Rule states that approximately 95% of the data items fall within 2 standard deviations of the mean. The figure shows that 95% of male adults have heights between 62 inches and 78 inches.

b. The 68-95-99.7 Rule states that approximately 95% of the data items fall within 2 standard deviations of the mean. Since the mean is 70 inches, the figure shows that half of the 95%, or 47.5%, of male adults have heights between 70 inches and 78 inches.

c. The 68-95-99.7 Rule states that approximately 68% of the data items fall within 1 standard deviation of the mean, thus 32% of the data falls outside this range. Half of the 32%, or 16% of male adults will have heights above 74 inches.

Normal Distribution of Male Adult Heights

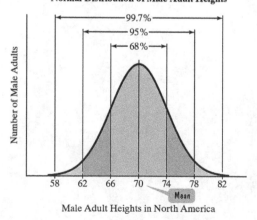

Male Adult Heights in North America

Your Turn (exercise #11, 15, 19, pg. 813):

Not everyone pays the same price for the same model of a car. The figure illustrates a normal distribution for the prices paid for a particular model of a new car. The mean is $17,000 and the standard deviation is $500.

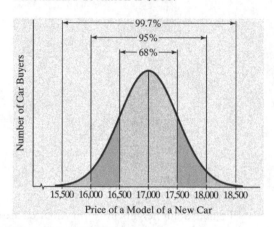

Price of a Model of a New Car

Use the 68-95-99.7 rule to find the percentage of buyers who paid

a. Between $16,500 and $17, 500

b. Between $16,000 and $17,000

c. More than $17,500

- **Objective 5 - Convert a data item to a z-score.**

Solved Problem:

The length of horse pregnancies from conception to birth is normally distributed with a mean of 336 days and a standard deviation of 3 days. Find the z-score for a horse pregnancy of

a. 342 days
b. 336 days
c. 333 days.

a.
$$z_{342} = \frac{\text{data item} - \text{mean}}{\text{standard deviation}} = \frac{342 - 336}{3} = \frac{6}{3} = 2$$

b.
$$z_{336} = \frac{\text{data item} - \text{mean}}{\text{standard deviation}} = \frac{336 - 336}{3} = \frac{0}{3} = 0$$

c.
$$z_{333} = \frac{\text{data item} - \text{mean}}{\text{standard deviation}} = \frac{333 - 336}{3} = \frac{-3}{3} = -1$$

Your Turn (exercise #33, 41, 45, pg. 813):

A set of data items is normally distributed with a mean of 60 and a standard deviation of 8. Convert each of the below data items to a z-score.

a. 68
b. 60
c. 48

248

- **Objective 6 - Understand percentiles and quartiles.**

Solved Problem:

A student scored in the 75th percentile on the SAT. What does this mean?

This means that 75% of the scores on the SAT are less than this student's score.

Your Turn (exercise #88, pg. 815):

Determine whether the below statement makes sense or does not make sense, and explain your reasoning.

I scored in the 50th percentile on a standardized test, so my score is the median.

- **Objective 7 - Use and interpret margins of error.**

Solved Problem:

Number of Books U.S. Adults Read per Year

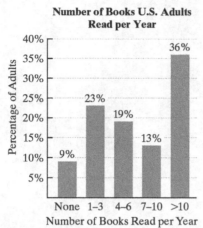

FIGURE 12.19
Source: Harris Poll of 2513 U.S. adults ages 18 and older conducted March 11 and 18, 2008

A Harris Poll of 2513 U.S. adults ages 18 and older asked the question How many books do you typically read in a year? The results of the poll are shown in **Figure 12.19** (p. 811).

a. Find the margin of error for this survey. Round to the nearest tenth of a percent.

b. Write a statement about the percentage of U.S. adults who read more than ten books per year.

a. The sample size is $n = 2513$. The margin of error is

$$\pm \frac{1}{\sqrt{n}} \times 100\% = \pm \frac{1}{\sqrt{2513}} \times 100\%$$
$$\approx \pm 0.020 \times 100\% = \pm 2.0\%$$

b. There is a 95% probability that the true population percentage lies between

the sample percent $- \frac{1}{\sqrt{n}} \times 100\%$

$= 36\% - 2.0\% = 34\%$ and

the sample percent $+ \frac{1}{\sqrt{n}} \times 100\%$

$= 36\% + 2.0\% = 38\%$
We can be 95% confident that between 34% and 38% of Americans read more than ten books per year.

Your Turn (exercise #69, pg. 814):

Using a random sample of 4000 TV households, Nielsen Media Research found that 60.2% watched the final episode of *M*A*S*H*.

a. Find the margin of error in this percent.

b. Write a statement about the percentage of TV households in the population that tuned into the final episode of *M*A*S*H*.

Section 12.5 Problem Solving with the Normal Distribution

- **Objective 1 - Solve applied problems involving normal distributions.**

Solved Problem:

The distribution of monthly charges for cellphone plans in the United States is approximately normal with a mean of $62 and a standard deviation of $18. What percentage of plans have charges that are less than $83.60? Refer to **Table 12.17** (p. 816).

$$z_{83.60} = \frac{\text{data item} - \text{mean}}{\text{standard deviation}} = \frac{83.60 - 62}{18} = 1.2$$

A *z*-score of 1.2 corresponds to a percentile of 88.49. Thus, 88.49% of plans have charges less than $83.60.

Your Turn (exercise #17, pg. 820):

Systolic blood pressure readings are normally distributed with a mean of 121 and a standard deviation of 15. A reading about 140 is considered to be high blood pressure. In the below exercises, begin by converting any given blood pressure reading or readings into *z*-scores. The use **Table 12.17** (p. 816) to find the percentage of people with blood pressure readings below 142.

Section 12.6 Scatter Plots, Correlation, and Regression Lines

- **Objective 2 - Interpret information given in a scatter plot.**

Solved Problem:

In a 1996 study involving obesity in mothers and daughters, researchers found a relationship between a high body-mass index for the girls and their mothers. (Body-mass index is a measure of weight relative to height. People with a high body-mass index are overweight or obese.) The correlation coefficient in this study was 0.51. Does this indicate a weak relationship between the body-mass index of daughters and the body-mass index of their mothers?

0.51 would indicate a moderate correlation between the two.

Your Turn (exercise #27, pg. 830-831):

Which scatter plot indicates a perfect negative correlation?

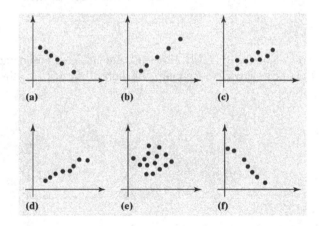

252

- **Objective 3 - Compute the correlation coefficient.**

Solved Problem:

The points in the scatter plot in **Figure 12.30** (p. 826) show the number of firearms per 100 persons and the number of deaths per 100,000 persons for the ten industrialized countries with the highest death rates. Use the data displayed by the voice balloons to determine the correlation coefficient between these variables. Round to two decimal places. What does the correlation coefficient indicate about the strength and direction of the relationship between firearms per 100 persons and deaths per 100,000 persons?

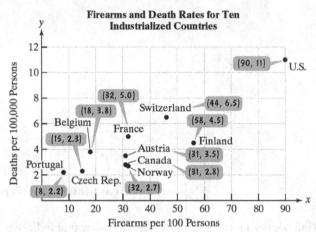

FIGURE 12.30
Source: International Action Network on Small Arms

This value for *r* is fairly close to 1 and indicates a moderately strong positive correlation. This means the higher the rate of firearm ownership, the higher the rate of deaths.

x	y	xy	x^2	y^2
8	2.2	17.6	64	4.84
15	2.3	34.5	225	5.29
18	3.8	68.4	324	14.44
31	2.8	86.8	961	7.84
31	3.5	108.5	961	12.25
32	2.7	86.4	1024	7.29
32	5.0	160	1024	25
44	6.5	286	1936	42.25
58	4.5	261	3364	20.25
90	11.0	990	8100	121

$$\sum x = 359 \quad \sum y = 44.3 \quad \sum xy = 2099.2 \quad \sum x^2 = 17,983 \quad \sum y^2 = 260.45$$

$$\left(\sum x\right)^2 = (359)^2 = 128,881 \text{ and } \left(\sum y\right)^2 = (44.3)^2 = 1962.49$$

$$r = \frac{10(2099.2) - (359)(44.3)}{\sqrt{10(17,983) - 128,881}\sqrt{10(260.45) - 1962.49}} = \frac{5088.3}{\sqrt{50949}\sqrt{642.01}} \approx 0.89$$

Copyright © 2015 Pearson Education, Inc.

253

Your Turn (exercise #35a, pg. 831):

Determine the correlation coefficient, rounded to two decimal places, between the percentage of people who won't try sushi and the percentage who don't approve of marriage equality.

Generation	Percentage Who	
	Won't Try Sushi x	Don't Approve of Marriage Equality y
Millennials	42	36
Gen X	52	49
Boomers	60	59
Silent/Greatest Generation	72	66

254

• **Objective 4 - Write the equation of the regression line.**

Solved Problem:

Use the data in **Figure 12.30** (p. 826) of Check Point 2 on page 826 to find the equation of the regression line. Round m and b to one decimal place. Then use the equation to project the number of deaths per 100,000 persons in a country with 80 firearms per 100 persons.

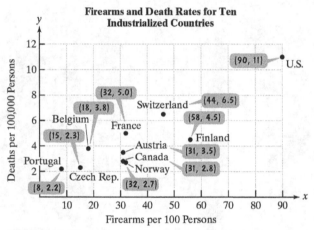

FIGURE 12.30
Source: International Action Network on Small Arms

$$m = \frac{10(2099.2) - (359)(44.3)}{10(17,983) - 128,881} = \frac{5088.3}{50949} \approx 0.1$$

$$b = \frac{44.3 - (0.1)(359)}{10} = \frac{8.4}{10} \approx 0.8$$

The equation of the regression line is $y = 0.1x + 0.8$.

The predicted rate in a country with 80 firearms per 100 persons can be found by substituting 80 for x.

$$y = 0.1x + 0.8$$
$$= 0.1(80) + 0.8$$
$$= 8.8$$

The death rate would be 8.8 per 100,000 people.

Copyright © 2015 Pearson Education, Inc.

Your Turn (exercise #35c, d, pg. 831):

| Generation | Percentage Who | |
	Won't Try Sushi x	Don't Approve of Marriage Equality y
Millennials	42	36
Gen X	52	49
Boomers	60	59
Silent/Greatest Generation	72	66

a. Find the equation of the regression line for the percentage who won't try sushi and the percentage who don't approve of marriage equality. Round m and b to two decimal places.

b. What percentage of people, to the nearest percent, can we anticipate do not approve of marriage equality in a generation where 30% won't try sushi?

256

- **Objective 5 - Use a sample's correlation coefficient to determine whether there is a correlation in the population.**

Solved Problem:

If you worked Check Point 2 correctly, you should have found that $r \approx 0.89$ for $n = 10$. Can you conclude that there is a positive correlation for all industrialized countries between firearms per 100 persons and deaths per 100,000 persons?

Yes, $|r| = 0.89$. Since $0.89 > 0.632$ and 0.765 (using **Table 12.19,** p. 828), we may conclude that a correlation does exist.

Your Turn (exercise #39, pg. 831):

The correlation coefficient, r, is given for a sample of n data points. Use the $\alpha = 0.05$ column in **Table 12.19** (pg. 828) to determine whether or not we may conclude that a correlation does exist in the population.

$n = 20, r = 0.5$

Chapter 13
Voting and Apportionment

Section 13.1 Voting Methods

- **Objective 1 - Understand and use preference tables.**

Solved Problem:

Four candidates are running for student body president: Alan (A), Bonnie (B), Carlos (C), and Samir (S). The students were asked to rank all the candidates in order of preference. Table 13.3 (p. 842) shows the preference table for this election.

a. How many students voted in the election?

b. How many students selected the candidates in this order: B, S, A, C?

c. How many students selected Samir (S) as their first choice for student body president?

TABLE 13.3 Preference Table for the Election of Student Body President				
Number of Votes	**2100**	**1305**	**765**	**40**
First Choice	S	A	S	B
Second Choice	A	S	A	S
Third Choice	B	B	C	A
Fourth Choice	C	C	B	C

a. We find the number of people who voted in the election by adding the numbers in the row labeled Number of Votes: 2100 + 1305 + 765 + 40 = 4210. Thus, 4210 people voted in the election.

b. We find how many people selected the candidates in the order B, S, A, C by referring to the fourth column of letters in the preference table. Above this column is the number 40. Thus, 40 people voted in the order B, S, A, C.

c. We find the number of people who selected S as their first choice by reading across the row that says First Choice: 2100 + 765 = 2865.

Thus, 2865 students selected S (Samir) as their first choice for student body president

258

Your Turn (exercise #5a, b, c, pg. 850):

Your class is given the option of choosing a day for the final exam. The students in the class are asked to rank the three available days, Monday (M), Wednesday (W), and Friday (F). The results of the election are shown in the following preference table.

Number of Votes	14	8	3	1
First Choice	F	F	W	M
Second Choice	W	M	F	W
Third Choice	M	W	M	F

a. How many students voted in the election?

b. How many students selected the days in this order: F, M, W?

c. How many students selected Friday as their first choice for the final?

- **Objective 2 - Use the plurality method to determine an election's winner.**

Solved Problem:

Table 13.2 on page 841 shows the preference table for the four candidates running for mayor of Smallville: Antonio (A), Bob (B), Carmen (C), and Donna (D). Who is declared the winner using the plurality method?

TABLE 13.2 Preference Table for the Smallville Mayoral Election				
Number of Votes	130	120	100	150
First Choice	A	D	D	C
Second Choice	B	B	B	B
Third Choice	C	C	A	A
Fourth Choice	D	A	C	D

The candidate with the most first-place votes is the winner. When using Table **13.2**, it is only necessary to look at the row which indicates the number of first-place votes. This indicates that A (Antonio) gets 130 first-place votes, C (Carmen) gets 150 first-place votes, and D (Donna) gets $120 + 100 = 220$ first-place votes. Thus Donna is declared the winner using the plurality method.

Your Turn (exercise #27, pg. 851):

Voters are asked to rank four brands of soup: A, B, C, and D. The votes are summarized in the following preference table.

Number of Votes	34	30	6	2
First Choice	A	B	C	D
Second Choice	B	C	D	B
Third Choice	C	D	B	C
Fourth Choice	D	A	A	A

Determine the winner using the plurality method.

- **Objective 3 - Use the Borda count method to determine an election's winner.**

Solved Problem:

Table 13.2 on page 841 shows the preference table for the four candidates running for mayor of Smallville: Antonio (A), Bob (B), Carmen (C), and Donna (D). Who is declared the winner using the Borda count method?

TABLE 13.2 Preference Table for the Smallville Mayoral Election				
Number of Votes	**130**	**120**	**100**	**150**
First Choice	A	D	D	C
Second Choice	B	B	B	B
Third Choice	C	C	A	A
Fourth Choice	D	A	C	D

Because there are four candidates, a first-place vote is worth 4 points, a second-place vote is worth 3 points, a third-place vote is worth 2 points, and a fourth-place vote is worth 1 point. We show the points produced by the votes in the preference table below.

Now we read down each column and total the points for each candidate separately.

A gets 520 + 120 + 200 + 300 = 1140 points
B gets 390 + 360 + 300 + 450 = 1500 points
C gets 260 + 240 + 100 + 600 = 1200 points
D gets 130 + 480 + 400 + 150 = 1160 points

Because B (Bob) has received the most points, he is the winner and the new mayor of Smallville.

Number of Votes	130	120	100	150
First Choice: 4 points	A: $130 \times 4 = 520$ pts	D: $120 \times 4 = 480$ pts	D: $100 \times 4 = 400$ pts	C: $150 \times 4 = 600$ pts
Second Choice: 3 points	B: $130 \times 3 = 390$ pts	B: $120 \times 3 = 360$ pts	B: $100 \times 3 = 300$ pts	B: $150 \times 3 = 450$ pts
Third Choice: 2 points	C: $130 \times 2 = 260$ pts	C: $120 \times 2 = 240$ pts	A: $100 \times 2 = 200$ pts	A: $150 \times 2 = 300$ pts
Fourth Choice: 1 point	D: $130 \times 1 = 130$ pts	A: $120 \times 1 = 120$ pts	C: $100 \times 1 = 100$ pts	D: $150 \times 1 = 150$ pts

Your Turn (exercise #28, pg. 851):

Voters are asked to rank four brands of soup: A, B, C, and D. The votes are summarized in the following preference table.

Number of Votes	34	30	6	2
First Choice	A	B	C	D
Second Choice	B	C	D	B
Third Choice	C	D	B	C
Fourth Choice	D	A	A	A

Determine the winner using the Borda count method.

- **Objective 4 - Use the plurality-with elimination method to determine an election's winner.**

Solved Problem:

Table 13.2 on page 841 shows the preference table for the four candidates running for mayor of Smallville: Antonio (A), Bob (B), Carmen (C), and Donna (D). Who is declared the winner using the plurality-with-elimination method?

TABLE 13.2 Preference Table for the Smallville Mayoral Election				
Number of Votes	**130**	**120**	**100**	**150**
First Choice	A	D	D	C
Second Choice	B	B	B	B
Third Choice	C	C	A	A
Fourth Choice	D	A	C	D

There are 130 + 120 + 100 + 150, or 500, people voting. In order to receive a majority, a candidate must receive more than 50% of the votes, meaning more than 250 votes. The number of first-place votes for each candidate is A (Antonio) = 130 B (Bob) = 0 C (Carmen) = 150 D (Donna) = 220

We see that no candidate receives a majority of first-place votes. Because Bob received the fewest first-place votes, he is eliminated in the next round. We construct a new preference table in which B is removed. Each candidate below B moves up one place, while the positions of candidates above B remain unchanged.

Number of Votes	130	120	100	150
First Choice	A	D	D	C
Second Choice	C	C	A	A
Third Choice	D	A	C	D

The number of first-place votes for each candidate is now A (Antonio) = 130; C (Carmen) = 150; D (Donna) = 220 No candidate receives a majority of first-place votes. Because Antonio received the fewest first-place votes, he is eliminated in the next round.

Number of Votes	130	120	100	150
First Choice	C	D	D	C
Second Choice	D	C	C	D

The number of first-place votes for each candidate is now C (Carmen) = 280; D (Donna) = 220 Because Carmen has received the majority of first-place votes, she is the winner and the new mayor of Smallville.

Your Turn (exercise #29, pg. 851):

Voters are asked to rank four brands of soup: A, B, C, and D. The votes are summarized in the following preference table.

Number of Votes	34	30	6	2
First Choice	A	B	C	D
Second Choice	B	C	D	B
Third Choice	C	D	B	C
Fourth Choice	D	A	A	A

Determine the winner using the plurality-with-elimination method.

264

- **Objective 5 - Use the pairwise comparison method to determine an election's winner.**

Solved Problem:

Table 13.2 on page 841 shows the preference table for the four candidates running for mayor of Smallville: Antonio (A), Bob (B), Carmen (C), and Donna (D). Make 6 comparisons (A vs. B, A vs. C, A vs. D, B vs. C, B vs. D, and C vs. D) and determine the winner using the pairwise comparison method.

TABLE 13.2 Preference Table for the Smallville Mayoral Election				
Number of Votes	**130**	**120**	**100**	**150**
First Choice	A	D	D	C
Second Choice	B	B	B	B
Third Choice	C	C	A	A
Fourth Choice	D	A	C	D

A vs. B

130 voters prefer A to B.
120 + 100 + 150 = 370 voters prefer B to A.
Conclusion: B wins this comparison and gets one point.

130	120	100	150
A	D	D	C
B	**B**	**B**	**B**
C	C	A	A
D	A	C	D

A vs. C

130 + 100 = 230 voters prefer A to C.
120 + 150 = 270 voters prefer C to A.
Conclusion: C wins this comparison and gets one point.

130	120	100	150
A	D	D	**C**
B	B	B	B
C	**C**	A	A
D	A	C	D

A vs. D

130 + 150 = 280 voters prefer A to D.
120 + 100 = 220 voters prefer D to A.
Conclusion: A wins this comparison and gets one point.

130	120	100	150
A	D	D	C
B	B	B	B
C	C	A	**A**
D	A	C	*D*

B vs. C

130	120	100	150
A	D	D	C
B	**B**	**B**	*B*
C	*C*	A	A
D	A	*C*	D

130 + 120 + 100 = 350 voters prefer B to C.
150 voters prefer C to B.
Conclusion: B wins this comparison and gets one point.

B vs. D

130	120	100	150
A	D	D	C
B	*B*	*B*	**B**
C	C	A	A
D	A	C	*D*

130 + 150 = 280 voters prefer B to D.
120 + 100 = 220 voters prefer D to B.
Conclusion: B wins this comparison and gets one point.

C vs. D

130	120	100	150
A	D	D	C
B	B	B	B
C	*C*	A	A
D	A	C	*D*

130 + 150 = 280 voters prefer C to D.
120 + 100 = 220 voters prefer D to C.
Conclusion: C wins this comparison and gets one point.

We now use each of the six conclusions and add points for the six comparisons.
A gets 1 point.
B gets 1 + 1 + 1 = 3 points.
C gets 1 + 1 = 2 points.

After all comparisons have been made, the candidate receiving the most points is B (Bob). He is the winner and the new mayor of Smallville.

Your Turn (exercise #30, pg. 851):

Voters are asked to rank four brands of soup: A, B, C, and D. The votes are summarized in the following preference table.

Number of Votes	34	30	6	2
First Choice	A	B	C	D
Second Choice	B	C	D	B
Third Choice	C	D	B	C
Fourth Choice	D	A	A	A

Determine the winner using the pairwise comparison method.

Section 13.2 Flaws of Voting Methods

- **Objective 1 - Use the majority criterion to determine a voting system's fairness.**

Solved Problem:

The 14 members of the school board must hire a new principal. The four finalists for the job, A, B, C, and D, are ranked by the 14 members. The preference table is shown in **Table 13.10** (p. 854). The board members agree to use the Borda count method to determine the winner.

a. Which candidate has a majority of first-place votes?
b. Which candidate is declared the new principal using the Borda count method?

TABLE 13.10 Preference Table for Selecting a New Principal				
Number of Votes	6	4	2	2
First Choice	A	B	B	A
Second Choice	B	C	D	B
Third Choice	C	D	C	D
Fourth Choice	D	A	A	C

a. There are 14 first-place votes. A candidate with more than half of these receives a majority. The first-choice row shows that candidate A received 8 first-place votes. Thus, candidate A has a majority of first-place votes.
b. Using the Borda count method with four candidates, a first-place vote is worth 4 points, a second-place vote is worth 3 points, a third place vote is worth 2 points, and a fourth-place vote is worth 1 point.

Now we read down the columns and total the points for each candidate.

A gets $24 + 4 + 2 + 8 = 38$ points.
B gets $18 + 16 + 8 + 6 = 48$ points.
C gets $12 + 12 + 4 + 2 = 30$ points.
D gets $6 + 8 + 6 + 4 = 24$ points.

Because candidate B has received the most points, candidate B is declared the new principal using the Borda count method.\

Number of Votes	6	4	2	2
First Choice: 4 points	A: $6 \times 4 = 24$ pts	B: $4 \times 4 = 16$ pts	B: $2 \times 4 = 8$ pts	A: $2 \times 4 = 8$ pts
Second Choice: 3 points	B: $6 \times 3 = 18$ pts	C: $4 \times 3 = 12$ pts	D: $2 \times 3 = 6$ pts	B: $2 \times 3 = 6$ pts
Third Choice: 2 points	C: $6 \times 2 = 12$ pts	D: $4 \times 2 = 8$ pts	C: $2 \times 2 = 4$ pts	D: $2 \times 2 = 4$ pts
Fourth Choice: 1 point	D: $6 \times 1 = 6$ pts	A: $4 \times 1 = 4$ pts	A: $2 \times 1 = 2$ pts	C: $2 \times 1 = 2$ pts

Your Turn (exercise #1, pg. 860):

Voters in a small town are considering four proposals, A, B, C, and D, for the design of affordable housing. The winning design is to be determined by the Borda count method. The preference table for the election is shown.

Number of Votes	300	120	90	60
First Choice	D	C	C	A
Second Choice	A	A	A	D
Third Choice	B	B	D	B
Fourth Choice	C	D	B	C

a. Which design has a majority of first-place votes?

b. Using the Borda count method, which design will be used for the affordable housing?

c. Is the majority criterion satisfied? Explain your answer.

268

- **Objective 2 - Use the head-to-head criterion to determine a voting system's fairness.**

Seven people are asked to listen to and rate three different pairs of stereo speakers, A, B, and C. The results are summarized in **Table 13.12** (p. 855).

    a.   Which brand is favored over all others using a head-to-head comparison?
    b.   Which brand wins the listening test using the plurality method?

TABLE 13.12 Preference Table for Three Pairs of Stereo Speakers

Number of Votes	3	2	2
First Choice	A	B	C
Second Choice	B	A	B
Third Choice	C	C	A

a.   We begin by comparing A and B. A is favored over B in column 1, giving A 3 votes. B is favored over A in columns 2 and 3, giving B 2 + 2, or 4, votes. Thus, B is favored when compared to A.

    Now we compare B to C. B is favored over C in columns 1 and 2, giving B 3 + 2, or 5, votes. C is favored over B in column 3, giving C 2 votes. Thus, B is favored when compared to C.
    We see that B is favored over both A and C using a head-to-head comparison.

b.   Using the plurality method, the brand with the most first-place votes is the winner. In the row indicating first choice, A received 3 votes, B received 2 votes, and C received 2 votes. A wins using the plurality method.

Your Turn (exercise #3, pg. 860):

MTV's *Real World* is considering three cities for its new season: Amsterdam (A), Rio de Janeiro (R), or Vancouver (V). Programming executives and the show's production team vote to decide where the new season will be taped. The winning city is to be determined by the plurality method. The preference table for the election is shown below.

Number of Votes	12	9	4	4
First Choice	A	V	V	R
Second Choice	R	R	A	A
Third Choice	V	A	R	V

a. Which city is favored over all others using a head-to-head comparison?

b. Which city wins the vote using the plurality method?

c. Is the head-to-head criterion satisfied? Explain your answer.

- **Objective 3 - Use the monotonicity criterion to determine a voting system's fairness.**

Solved Problem:

An election with 120 voters and three candidates, A, B, and C, is to be decided using the plurality-with-elimination method. **Table 13.17** (p. 857) shows the results of a straw poll. After the straw poll, 12 voters all changed their ballots to make candidate A their first choice. The results of the second election are given in **Table 13.18** (p. 857).

a. Using the plurality-with-elimination method, which candidate wins the first election?

b. Using the plurality-with-elimination method, which candidate wins the second election?

c. Does this violate the monotonicity criterion? Explain your answer.

TABLE 13.17 Preference Table for the Straw Vote

Number of Votes	42	34	28	16
First Choice	A	C	B	B
Second Choice	B	A	C	A
Third Choice	C	B	A	C

TABLE 13.18 Preference Table for the Second Election

Number of Votes	54	34	28	4
First Choice	A	C	B	B
Second Choice	B	A	C	A
Third Choice	C	B	A	C

a. There are 120 people voting. No candidate initially receives more than 60 votes. Because C receives the fewest first-place votes, C is eliminated in the next round. The new preference table is

Number of Votes	42	34	28	16
First Choice	A	A	B	B
Second Choice	B	B	A	A

Because A has received a majority of first-place votes, A is the winner of the straw poll.

b. No candidate initially receives more than 60 votes. Because B receives the fewest first-place votes, B is eliminated in the next round. The new preference table is

Number of Votes	54	34	28	4
First Choice	A	C	C	A
Second Choice	C	A	A	C

Because C has received a majority of first-place votes, C is the winner of the second election.

c.   A won the first election. A then gained additional support with the 12 voters who changed their ballots to make A their first choice. A lost the second election. This violates the monotonicity criterion.

Your Turn (exercise #7, pg. 861):

The following preference table gives the results of a straw vote among three candidates, A, B, and C.

Number of Votes	10	8	7	4
First Choice	C	B	A	A
Second Choice	A	C	B	C
Third Choice	B	A	C	B

a.   Using the plurality-with-elimination method, which candidate wins the straw vote?

b.   In the actual election, the four voters in the last column who voted A, C, B, in that order, change their votes to C, A, B. Using the plurality-with-elimination method, which candidate wins the actual election?

c.   Is the monotonicity criterion satisfied? Explain your answer.

- **Objective 4 - Use the irrelevant alternatives criterion to determine a voting system's fairness.**

Solved Problem:

Four candidates, A, B, C, and D, are running for mayor. The election results are shown in **Table 13.21** (p. 858).

a. Using the pairwise comparison method, who wins this election?

b. Prior to the announcement of the election results, candidates B and C both withdraw from the running. Using the pairwise comparison method, which candidate is declared mayor with B and C eliminated from the preference table?

c. Does this violate the irrelevant alternatives criterion? Explain your answer.

TABLE 13.21 Preference Table for Mayor

Number of Votes	150	90	90	30
First Choice	A	C	D	D
Second Choice	B	B	A	A
Third Choice	C	D	C	B
Fourth Choice	D	A	B	C

a. Because there are 4 candidates, $n = 4$ and the number of comparisons we must make is
$$\frac{n(n-1)}{2} = \frac{4(4-1)}{2} = \frac{4 \cdot 3}{2} = \frac{12}{2} = 6$$
The following table shows the results of these 6 comparisons.

Comparison	Vote Results	Conclusion
A vs. B	270 voters prefer A to B. 90 voters prefer B to A.	A wins and gets 1 point.
A vs. C	270 voters prefer A to C. 90 voters prefer C to A.	A wins and gets 1 point.
A vs. D	150 voters prefer A to D. 210 voters prefer D to A.	D wins and gets 1 point.
B vs. C	180 voters prefer B to C. 180 voters prefer C to B.	B and C tie. Each gets $\frac{1}{2}$ point.
B vs. D	240 voters prefer B to D. 120 voters prefer D to B.	B wins and gets 1 point.
C vs. D	240 voters prefer C to D. 120 voters prefer D to C.	C wins and gets 1 point.

Thus A gets 2 points, B gets $1\frac{1}{2}$ points, C gets $1\frac{1}{2}$ points, and D gets 1 point. Therefore A is the winner.

b. After B and C withdraw, there is a new preference table:

Number of Votes	150	90	90	30
First Choice	A	D	D	D
Second Choice	D	A	A	A

Using the pairwise comparison test with 2 candidates, there is only one comparison to make namely A vs. D. 150 voters prefer A to D, and 210 voters prefer D to A. D gets 1 point, A gets 0 points, and D wins the election.

c.   The first election count produced A as the winner. The removal of B and C from the ballots produced D as the winner. This violates the irrelevant alternatives criterion.

Your Turn (exercise #9, pg. 861):

Members of the Student Activity Committee at a college are considering three film directors to speak at a campus arts festival. Ron Howard (H), Spike Lee (L), and Steven Spielberg (S). Committee members vote for their preferred speaker. The winner is to be selected by the pairwise comparison method. The preference table for the election is shown below.

Number of Votes	10	8	5
First Choice	H	L	S
Second Choice	S	S	L
Third Choice	L	H	H

a.   Using the pairwise comparison method, who is selected as the speaker?

b.   Prior to the announcement of the speaker, Ron Howard informs the committee that he will not be able to participate due to other commitments. Construct a new preference table for the election with H eliminated. Using the new table and the pairwise comparison method, who is selected as the speaker?

Is the irrelevant alternatives criterion satisfied? Explain your answer.

274

Section 13.3 Apportionment Methods

- **Objective 1 - Find standard divisors and standard quotas.**

Solved Problem:

The Republic of Amador is composed of five states, A, B, C, D, and E. According to the country's constitution, the congress will have 200 seats, divided among the five states according to their respective populations. **Table 13.27** (p. 863) shows each state's population.

a. Find the standard divisor.

b. Find the standard quota for each state and complete **Table 13.27**.

TABLE 13.27 Population of Amador by State						
State	A	B	C	D	E	Total
Population (in thousands)	1112	1118	1320	1515	4935	10,000
Standard Quota						

a. Standard divisor $= \dfrac{\text{total population}}{\text{number of allocated items}} = \dfrac{10,000}{200} = 50$

b. Standard quota for state A $=$
$\dfrac{\text{population of state A}}{\text{standard divisor}} = \dfrac{1112}{50} = 22.24$

Standard quota for state B $=$
$\dfrac{\text{population of state B}}{\text{standard divisor}} = \dfrac{1118}{50} = 22.36$

Standard quota for state C $=$
$\dfrac{\text{population of state C}}{\text{standard divisor}} = \dfrac{1320}{50} = 26.4$

Standard quota for state D $=$
$\dfrac{\text{population of state D}}{\text{standard divisor}} = \dfrac{1515}{50} = 30.3$

Standard quota for state E $=$
$\dfrac{\text{population of state E}}{\text{standard divisor}} = \dfrac{4935}{50} = 98.7$

Copyright © 2015 Pearson Education, Inc.

Table 13.27	Population of Amador by State					
State	A	B	C	D	E	Total
Population (in thousands)	1112	1118	1320	1515	4935	10,000
Standard quota	22.24	22.36	26.4	30.3	98.7	200

Your Turn (exercise #1a, b, pg. 874):

A small country is comprised of four states, A, B, C, and D. The population of each state, in thousands, is given in the following table.

State	A	B	C	D	Total
Population (in thousands)	138	266	534	662	1600

According to the country's constitution, the congress will have 80 seats, divided among the four states according to their respective populations.

a. Find the standard divisor, in thousands. How many people are there for each seat in congress?

b. Find each state's standard quota.

276

- **Objective 3 - Use the monotonicity criterion to determine a voting system's fairness.**

Solved Problem:

The Republic of Amador is composed of five states, A, B, C, D, and E. According to the country's constitution, the congress will have 200 seats, divided among the five states according to their respective populations. **Table 13.27** (p. 863) shows each state's population.

TABLE 13.27 Population of Amador by State						
State	A	B	C	D	E	Total
Population (in thousands)	1112	1118	1320	1515	4935	10,000
Standard Quota						

Use Hamilton's method to apportion the 200 congressional seats.

State	Population (in thousands)	Standard Quota	Lower Quota	Fractional Part	Surplus	Final Apportionment
A	1112	22.24	22	0.24		22
B	1118	22.36	22	0.36		22
C	1320	26.4	26	0.4 (next largest)	1	27
D	1515	30.3	30	0.3		30
E	4935	98.7	98	0.7 (largest)	1	99
Total	10,000	200	198			200

Your Turn (exercise #3, pg. 874):

A small country is comprised of four states, A, B, C, and D. The population of each state, in thousands, is given in the following table.

State	A	B	C	D	Total
Population (in thousands)	138	266	534	662	1600

According to the country's constitution, the congress will have 80 seats, divided among the four states according to their respective populations. Use Hamilton's method to find each state's apportionment of congressional seats.

- **Objective 5 - Use Jefferson's method.**

Solved Problem:

The Republic of Amador is composed of five states, A, B, C, D, and E. According to the country's constitution, the congress will have 200 seats, divided among the five states according to their respective populations. **Table 13.27** (p. 863) shows each state's population.

TABLE 13.27 Population of Amador by State

State	A	B	C	D	E	Total
Population (in thousands)	1112	1118	1320	1515	4935	10,000
Standard Quota						

Use Jefferson's method with $d = 49.3$ to apportion the 200 congressional seats.

State	Population (in thousands)	Modified Quota (using $d = 49.3$)	Modified Lower Quota	Final Apportionment
A	1112	22.56	22	22
B	1118	22.68	22	22
C	1320	26.77	26	26
D	1515	30.73	30	30
E	4935	100.10	100	100
Total	10,000		200	200

Your Turn (exercise #39, pg. 875):

An HMO has 150 doctors to be apportioned among four clinics. The HMO decides to apportion the doctors based on the average weekly patient load for each clinic, given in the following table. Use Jefferson's method to apportion the 150 doctors. (*Hint*: Find the standard divisor. A modified divisor that is less than this standard divisor will work).

Clinic	A	B	C	D
Average Weekly Patient Load	1714	5460	2440	5386

278

- **Objective 6 - Use Adams's method.**

Solved Problem:

The Republic of Amador is composed of five states, A, B, C, D, and E. According to the country's constitution, the congress will have 200 seats, divided among the five states according to their respective populations. **Table 13.27** (p. 863) shows each state's population.

TABLE 13.27 Population of Amador by State						
State	A	B	C	D	E	Total
Population (in thousands)	1112	1118	1320	1515	4935	10,000
Standard Quota						

Use Adams's method to apportion the 200 congressional seats. In guessing at a value for d, begin with $d = 50.5$. If necessary, modify this value as we did in Example 4 (p. 870-871).

State	Population (in thousands)	Modified Quota (using $d = 50.5$)	Modified Upper Quota
A	1112	22.02	23
B	1118	22.14	23
C	1320	26.14	27
D	1515	30	30
E	4935	97.72	98
Total	10,000		201

This sum should be 200, not 201.

State	Population (in thousands)	Modified Quota (using $d = 50.6$)	Modified Upper Quota	Final Apportionment
A	1112	21.98	22	22
B	1118	22.09	23	23
C	1320	26.09	27	27
D	1515	29.94	30	30
E	4935	97.53	98	98
Total	10,000		200	200

Your Turn (exercise #11, pg. 875):

The police department in a large city has 180 new officers to be apportioned among six high-crime precincts. Crimes by precinct are shown in the following table. Use Adams's method with $d = 16$ to apportion the new officers among the precincts.

Precinct	A	B	C	D	E	F
Crimes	446	526	835	227	338	456

- **Objective 7 - Use Webster's method**.

Solved Problem:

The Republic of Amador is composed of five states, A, B, C, D, and E. According to the country's constitution, the congress will have 200 seats, divided among the five states according to their respective populations. **Table 13.27** (p. 863) shows each state's population.

TABLE 13.27 Population of Amador by State						
State	**A**	**B**	**C**	**D**	**E**	**Total**
Population (in thousands)	1112	1118	1320	1515	4935	10,000
Standard Quota						

Use Webster's method with $d = 49.8$ to apportion the 200 congressional seats.

State	Population (in thousands)	Modified Quota (using $d = 49.8$)	Modified Rounded Quota
A	1112	22.33	22
B	1118	22.45	22
C	1320	26.51	27
D	1515	30.42	30
E	4935	99.10	99
Total	10,000		200

Your Turn (exercise #15, pg. 875):

Twenty sections of bilingual math courses, taught in both English and Spanish, are to be offered in introductory algebra, intermediate algebra, and liberal arts math. The preregistration figures for the number of students planning to enroll in these bilingual sections are given in the following table. Use Webster's method with $d = 29.6$ to determine how many bilingual sections of each course should be offered.

Course	Introductory Algebra	Intermediate Algebra	Liberal Arts Math
Enrollment	130	282	188

Section 13.4 Flaws of Apportionment Methods

- **Objective 1 - Understand and illustrate the Alabama paradox.**

Solved Problem:

Table 13.42 (p. 879) shows the populations of the four states in a country with a population of 20,000. Use Hamilton's method to show that the Alabama paradox occurs if the number of seats in congress is increased from 99 to 100.

TABLE 13.42					
State	A	B	C	D	Total
Population	2060	2080	7730	8130	20,000

We begin with 99 seats in the Congress. First we compute the standard divisor:

$$\text{Standard divisor} = \frac{\text{total population}}{\text{number of allocated items}} = \frac{20{,}000}{99} = 202.02$$

Using this value, make a table showing apportionment using Hamilton's method.

State	Population	Standard Quota	Lower Quota	Fractional Part	Surplus Seats	Final Apportionment
A	2060	10.20	10	0.20		10
B	2080	10.30	10	0.30	1	11
C	7730	38.26	38	0.26		38
D	8130	40.24	40	0.24		40
Total	20,000	99	98			99

Now let's see what happens with 100 seats in Congress. First we compute the standard divisor:

$$\text{Standard divisor} = \frac{\text{total population}}{\text{number of allocated items}} = \frac{20{,}000}{100} = 200$$

Using this value, make a table showing apportionment using Hamilton's method.

State	Population	Standard Quota	Lower Quota	Fractional Part	Surplus Seats	Final Apportionment
A	2060	10.3	10	0.3		10
B	2080	10.4	10	0.4		10
C	7730	38.65	38	0.65	1	39
D	8130	40.65	40	0.65	1	41
Total	20,000	100	98			100

The final apportionments are summarized in the following table.

State	Apportionment with 99 seats	Apportionment with 100 seats
A	10	10
B	11	10
C	38	39
D	40	41

When the number of seats increased from 99 to 100, B's apportionment decreased from 11 to 10.

Your Turn (exercise #1, pg. 885):

The mathematics department has 30 teaching assistants to be divided among three courses, according to their specific enrollments. The table shows the courses and the number of students enrolled in each courses.

Course	College Algebra	Statistics	Liberal Arts Math	Total
Enrollment	978	500	322	1800

a.   Apportion the teaching assistants using Hamilton's method.

b.   Use Hamilton's method to determine if the Alabama paradox occurs if the number of teaching assistants is increased from 30 to 31. Explain your answer.

- **Objective 2 - Understand and illustrate the population paradox.**

Solved Problem:

A small country has 100 seats in the congress, divided among the three states according to their respective populations. **Table 13.47** (p. 881) shows each state's population before and after the country's population increase.

a.   Use Hamilton's method to apportion the 100 congressional seats using the original population.

b.   Find the percent increase in the populations of states A and B. (State C did not have any change in population.)

c.   Use Hamilton's method to apportion the 100 congressional seats using the new population. Show that the population paradox occurs.

TABLE 13.47				
State	A	B	C	Total
Original Population	19,110	39,090	141,800	200,000
New Population	19,302	39,480	141,800	200,582

a.   We use Hamilton's method to find the apportionment for each state with its original population. First we compute the standard divisor.

$$\text{Standard divisor} = \frac{\text{total population}}{\text{number of allocated items}} = \frac{200,000}{100} = 2000$$

Using this value, we show the apportionment in the following table.

State	Original Population	Standard Quota	Lower Quota	Fractional Part	Surplus Seats	Final Apportionment
A	19,110	9.56	9	0.56	1	10
B	39,090	19.55	19	0.55		19
C	141,800	70.9	70	0.9	1	71
Total	200,000	100.01	98			100

284

b. The fraction for percent increase is the amount of increase divided by the original amount. The percent
increase in the population of each state is determined as follows.

State A: $\dfrac{19{,}302-19{,}110}{19{,}110} = \dfrac{192}{19{,}110} \approx 0.01005 = 1.005\%$

State B: $\dfrac{39{,}480-39{,}090}{39{,}090} = \dfrac{390}{39{,}090} \approx 0.00998 = 0.998\%$

State A is increasing at a rate of 1.005%. This is faster than State B, which is increasing at a rate of 0.998%.

c. We use Hamilton's method to find the apportionment for each state with its new population. First we
compute the standard divisor.

Standard divisor $= \dfrac{\text{total population}}{\text{number of allocated items}} = \dfrac{200{,}582}{100} = 2005.82$

Using this value, we show the apportionment in the following table.

State	New Population	Standard Quota	Lower Quota	Fractional Part	Surplus Seats	Final Apportionment
A	19,302	9.62	9	0.62		9
B	39,480	19.68	19	0.68	1	20
C	141,800	70.69	70	0.69	1	71
Total	200,582	99.99	98			100

The final apportionments are summarized in the following table.

State	Growth Rate	Original Apportionment	New Apportionment
A	1.005%	10	9
B	0.998%	19	20
C	0%	71	71

State A loses a seat to State B, even though the population of State A is increasing at a faster rate. This is an
example of the population paradox.

Your Turn (exercise #5, pg. 885):

A small country has 24 seats in the congress, divided among the three states according to their respective
populations. The table shows each state's population, in thousands, before and after the country's population
increase.

State	A	B	C	Total
Original Population (in thousands)	530	990	2240	3760
New Population (in thousands)	680	1250	2570	4500

a. Use Hamilton's method to apportion the 24 congressional seats using the original population.

b. Find the percent increase, to the nearest tenth of a percent, in the population of each state.

c. Use Hamilton's method to apportion the 24 congressional seats using the new population. Does the population paradox occur? Explain your answer.

286

- **Objective 3 - Understand and illustrate the new-states paradox.**

Solved Problem:

a. A school district has two high schools, East High, with an enrollment of 2574 students, and West High, with an enrollment of 9426 students. The school district has a counseling staff of 100 counselors. Use Hamilton's method to apportion the counselors to the two schools.

b. Suppose that a new high school, North High, with a population of 750 students, is added to the district. The school district decides to hire 6 new counselors for North High. Use Hamilton's method to show that the new-states paradox occurs when the counselors are reapportioned.

a. We use Hamilton's method to find the apportionment for each school. First we compute the standard divisor.

$$\text{Standard divisor} = \frac{\text{total population}}{\text{number of allocated items}} = \frac{12,000}{100} = 120$$

Using this value, we show the apportionment in the following table.

School	Enrollment	Standard Quota	Lower Quota	Fractional Part	Surplus	Final Apportionment
East High	2574	21.45	21	0.45		21
West High	9426	78.55	78	0.55	1	79
Total	12,000	100	99			100

b. Again we use Hamilton's method.

$$\text{Standard divisor} = \frac{\text{total population}}{\text{number of allocated items}} = \frac{12,750}{106} = 120.28$$

Using this value, we show the apportionment in the following table.

School	Enrollment	Standard Quota	Lower Quota	Fractional Part	Surplus	Final Apportionment
East High	2574	21.40	21	0.40	1	22
West High	9426	78.37	78	0.37		78
North High	750	6.24	6	0.24		6
Total	12,750	106.01	105			106

West High has lost a counselor to East High.

Your Turn (exercise #9, pg. 886):

A corporation has two branches, A and B. Each year, the company awards 100 promotions within its branches.
The table shows the number of employees in each branch.

Branch	A	B	Total
Employees	1045	8955	10,000

a.   Use Hamilton's method to apportion the promotions.

b.   Suppose that a third branch, C, with the number of employees shown in the table below, is added to the
      corporation. The company adds five new yearly promotions for branch C. Use Hamilton's method to
      determine if the new-states paradox occurs when the promotions are reapportioned.

Branch	A	B	C	Total
Employees	1045	8955	525	10,525

Chapter 14
Graph Theory

Section 14.1 Graphs, Paths, and Circuits

- **Objective 1 - Understand relationships in a graph.**

Solved Problem:

Explain why **Figures 14.4(a)** and **(b)** (p. 893) show equivalent graphs.

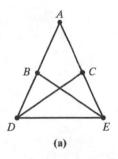
(a)

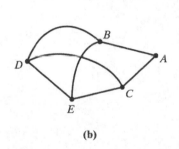
(b)

Graphs (a) and (b) both have vertices *A, B, C, D,* and *E*. Also, both graphs have edges *AB, AC, BD, BE, CD, CE,* and *DE*. Because the two graphs have the same number of vertices connected to each other in the same way, they are the same. In fact, graph (b) is just graph (a) rotated clockwise and bent out of shape.

Your Turn (exercise #7, pg. 900):

Explain why the two figures show equivalent graphs. Then draw a third equivalent graph.

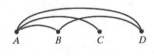

290

- **Objective 2 - Model relationships using graphs.**

Solved Problem:

The city of Metroville is located on both banks and three islands of the Metro River. **Figure 14.7** (p. 894) shows that the town's sections are connected by five bridges. Draw a graph that models the layout of Metroville.

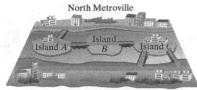

North Metroville

Island *A*   Island *B*   Island *C*

South Metroville   FIGURE 14.7

Draw points for the five land masses and label them *N, S, A, B,* and *C*. There is one bridge that connects North Metroville to Island A, so one edge is drawn connecting vertex *N* to vertex *A*. Similarly, one edge connects vertex *A* with vertex *B*, and one edge connects vertex *B* with vertex *C*. Since there are two bridges connecting Island C to South Metroville, two edges connect vertex *C* with vertex *S*.

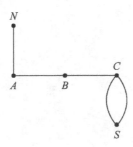

Your Turn (exercise #11, pg. 900):

Draw a graph that models the layout of the city shown in the map. Use vertices to represent the land masses and edges to represent the bridges.

The City of Gothamville:

North Gothamville

Gotham River

South Gothamville

292

- **Objective 3 - Understand and use the vocabulary of graph theory.**

Solved Problem:

List the pairs of adjacent vertices for the graph in **Figure 14.17** (p. 897).

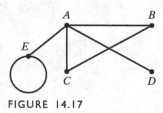

FIGURE 14.17

We systematically list which pairs of vertices are adjacent, working alphabetically. Thus, the adjacent vertices are *A* and *B*, *A* and *C*, *A* and *D*, *A* and *E*, *B* and *C*, and *E* and *E*.

Your Turn (exercise #25, pg. 901):

In the following graph, which vertices are adjacent to vertex *A*?

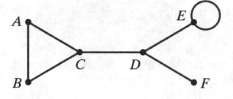

Section 14.2 Euler Paths and Euler Circuits

- **Objective 3 - Use Euler's Theorem.**

Solved Problem:

Refer to the graph in **Figure 14.26** (p. 905). Use trial and error to find an Euler circuit that starts and ends at *G*. Number the edges of the graph to indicate the circuit. Then use vertex letters separated by commas to name the circuit.

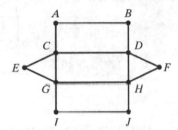

FIGURE 14.26

We count the number of edges at each vertex to determine if the vertex is odd or even. We see that the graph has no odd vertices. By the second statement in Euler's Theorem, the graph has at least one Euler circuit.

We use trial and error to find an Euler circuit that starts at *G*. The following figure shows a result.

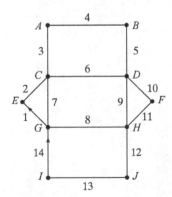

Using vertex letters to name the circuit, we write *G, E, C, A, B, D, C, G, H, D, F, H, J, I, G.*

Your Turn (exercise #9, pg. 910):

a. Explain why the graph has at least one Euler circuit.
b. Use trial and error or Fleury's Algorithm to find one such circuit.

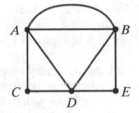

294

- **Objective 4 - Solve problems using Euler's Theorem.**

Solved Problem:

The floor plan of a four-room house and a graph that models the connecting relationships in the floor plan are shown in **Figure 14.34** (p. 907).

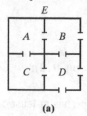
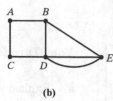

(a)   **(b)**

FIGURE 14.34 A floor plan and a graph that models its connecting relationships

a. Is it possible to find a path that uses each door exactly once?
b. If it is possible, use trial and error to show such a path on the graph in **Figure 14.34(b)** and the floor plan in **Figure 14.34(a)** (p. 907).

a. A walk through every room and the outside, using each door exactly once, means that we are looking for an Euler path or Euler circuit on the graph in Figure 14.34(b). This graph has exactly two odd vertices, namely *B* and *E*. By Euler's theorem, the graph has at least one Euler path, but no Euler circuit. It is possible to walk through every room and the outside, using each door exactly once. It is not possible to begin and end the walk in the same place.

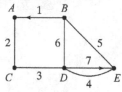

b. Euler's theorem tells us that a possible Euler path must start at one of the odd vertices and end at the other. We use trial and error to find such a path, starting at vertex *B* (room *B* in the floor plan), and ending at vertex *E* (outside in the floor plan). Possible paths follow.

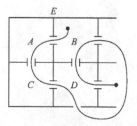

Your Turn (exercise #45, 46, pg. 912):

In the below exercises, we revisit the four-block neighborhood discussed in the previous section. Recall that a mail carrier parks her truck at the intersection shown in the figure and then walks to deliver mail to each of the houses. The streets on the outside of the neighborhood have houses on one side only. The interior streets have houses on both sides of the street. On these streets, the mail carrier must walk down the street twice, covering each side of the street separately. A graph that models the streets of the neighborhood walked by the mail carrier is shown.

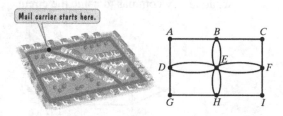

a. Use Euler's Theorem to explain why it is possible for the mail carrier to park at *B*, deliver mail to each house without retracing her route, and then return to *B*.

b. Use trial and error or Fleury's Algorithm to find an Euler circuit that starts and ends at vertex *B* on the graph that models the neighborhood.

- **Objective 5 - Use Fleury's Algorithm to find possible Euler paths and Euler circuits.**

Solved Problem:

Your Turn (exercise #33, pg. 911):

The graph in **Figure 14.37** (p. 908) has at least one Euler circuit. Find one by Fleury's Algorithm.

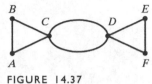

FIGURE 14.37

Use Fleury's Algorithm to find an Euler path.

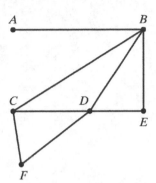

The graph has no odd vertices, so we can begin at any vertex. We choose vertex C as the starting point. From C we can travel to A, B, or D. We choose to travel to D.

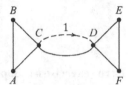

Now the remaining edge CD is a bridge, so we must travel to either E or F. We choose F.

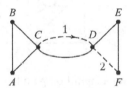

We have no choices for our next three steps, which are bridges. We must travel to E, then D, then C.

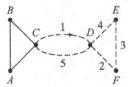

From C, we may travel to either A or B. We choose B. Then we must travel to A, then back to C.

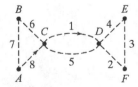

The above figure shows the completed Euler circuit. Written using the letters of the vertices, the path is C, D, F, E, D, C, B, A, C.

Section 14.3 Hamilton Paths and Hamilton Circuits

- **Objective 1 - Understand the definitions of Hamilton paths and Hamilton circuits.**

Solved Problem:

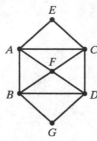

FIGURE 14.39

a. Find a Hamilton path that begins at vertex E for the graph in **Figure 14.39** (p. 915).

b. Find a Hamilton circuit that begins at vertex E for the graph in **Figure 14.39**.

a. A Hamilton path must pass through each vertex exactly once. The graph has many Hamilton paths. An example of such a path is E, C, D, G, B, A, F.

b. A Hamilton circuit must pass through every vertex exactly once and begin and end at the same vertex. The graph has many Hamilton circuits. An example of such a circuit is E, C, D, G, B, F, A, E.

Your Turn (exercise #1, 3, pg. 921):

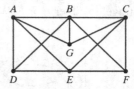

a. Find a Hamilton path that begins at A and ends at B.

b. Find a Hamilton path that begins as A, B, …

- **Objective 2 - Find the number of Hamilton circuits in a complete graph.**

Solved Problem:

Determine the number of Hamilton circuits in a complete graph with

a. three vertices
b. six vertices.
c. ten vertices.

In each case, we use the expression $(n-1)!$.
For three vertices, substitute 3 for n in the expression. For six and ten vertices, substitute 6 and 10, respectively, for n.

a. A complete graph with three vertices has
$(3-1)! = 2! = 2 \cdot 1 = 2$ Hamilton circuits.

b. A complete graph with six vertices has
$(6-1)! = 5! = 5 \cdot 4 \cdot 3 \cdot 2 \cdot 1 = 120$ Hamilton
circuits.

c. A complete graph with ten vertices has
$(10-1)! = 9! = 9 \cdot 8 \cdot 7 \cdot 6 \cdot 5 \cdot 4 \cdot 3 \cdot 2 \cdot 1 = 362,880$
Hamilton circuits.

Your Turn (exercise #15, 17, pg. 922):

Determine the number of Hamilton circuits in a complete graph with

a. Three vertices
b. Twelve vertices

298

- **Objective 3 - Understand and use weighted graphs.**

Solved Problem:

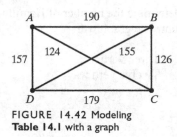

FIGURE 14.42 Modeling
Table 14.1 with a graph

Use the weighted graph in **Figure 14.42** (p. 917) to find the cost of the trip for the Hamilton circuit *A, C, B, D, A*.

The trip described by the Hamilton circuit *A, C, B, D, A* involves the sum of four costs:

$124 + $126 + $155 + $157 = $562.

Here, $124 is the cost of the trip from *A* to *C*; $126 is the cost from *C* to *B*; $155 is the cost from *B* to *D*; and $157 is the cost from *D* to *A*.

The total cost of the trip is $562.

Your Turn (exercise #21, pg. 922):

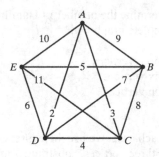

Find the total weight of the Hamilton circuit *A, B, C, E, D, A*.

- **Objective 4 - Use the Brute Force Method to solve traveling salesperson problems.**

Solved Problem:

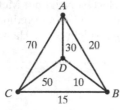

FIGURE 14.44

Use the Brute Force Method to find the optimal solution for the complete, weighted graph in **Figure 14.44** (p. 919). List Hamilton circuits as in **Table 14.2** (p. 918).

The graph has four vertices. Thus, using $(n-1)!$, there are $(4-1)! = 3! = 6$ possible Hamilton circuits. The 6 possible Hamilton circuits and their costs are shown.

The two Hamilton circuits having the lowest cost of $115 are *A, B, C, D, A* and *A, D, C, B, A*.

Your Turn (exercise #25-31, pg. 922):

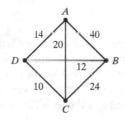

Find the total weight of each Hamilton circuit:

a.　A, B, C, D, A
b.　A, B, D, C, A
c.　A, C, B, D, A
d.　A, C, D, B, A
e.　A, D, B, C, A
f.　A, D, C, B, A

Use your answers from the above exercises and the Brute Force Method to find the optimal solution.

Hamilton Circuit	Sum of the Weights of the Edges	=	Total Cost
A, B, C, D, A	20 + 15 + 50 + 30	=	$115
A, B, D, C, A	20 + 10 + 50 + 70	=	$150
A, C, B, D, A	70 + 15 + 10 + 30	=	$125
A, C, D, B, A	70 + 50 + 10 + 20	=	$150
A, D, B, C, A	30 + 10 + 15 + 70	=	$125
A, D, C, B, A	30 + 50 + 15 + 20	=	$115

300

• **Objective 5 - Use the Nearest Neighbor Method to approximate solutions to traveling salesperson problems.**

Solved Problem:

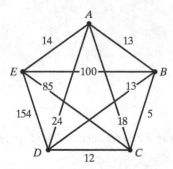

FIGURE 14.47

Use the Nearest Neighbor Method to approximate the optimal solution for the complete, weighted graph in **Figure 14.47** (p. 920). Begin the circuit at vertex A. What is the total weight of the resulting Hamilton circuit?

The Nearest Neighbor method is carried out as follows:

Start at A.

Choose the edge with the smallest weight: 13. Move along this edge to B.

From B, choose the edge with the smallest weight that does not lead to A: 5. Move along this edge to C.

From C, choose the edge with the smallest weight that does not lead to a city already visited: 12. Move along this edge to D.

From D, the only choice is to fly to E, the only city not yet visited: 154.

From E, close the circuit and return home to A: 14.

An approximate solution is the Hamilton circuit A, B, C, D, E, A. The total weight is:
$13 + 5 + 12 + 154 + 14 = 198$.

Your Turn (exercise #32, pg. 922):

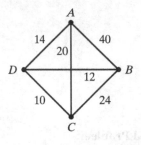

Use the Nearest Neighbor Method, with starting vertex A, to find an approximate solution. What is the total weight of the Hamilton circuit?

Section 14.3 Trees

- **Objective 1 - Understand the definition and properties of a tree.**

Solved Problem:

Which graph in **Figure 14.51** (p. 926) is a tree? Explain why the other two graphs shown are not trees.

The graph in Figure 14.51(c) is a tree. It is connected and has no circuits. There is only one path joining any two vertices. Every edge is a bridge; if removed, each edge would create a disconnected graph. Finally, the graph has 7 vertices and 7 – 1, or 6, edges.

The graph in Figure 14.51(a) is not a tree because it is disconnected. There are 7 vertices and only 5 edges, not the 6 edges required for a tree.

The graph in Figure 14.51(b) is not a tree because it has a circuit, namely A, B, C, D, A. There are 7 vertices and 7 edges, not the 6 edges required for a tree.

Your Turn (exercise #3, 5, 7, pg. 930):

Determine whether each graph is a tree. If the graph is not a tree, give a reason why.

a. Graph from #3

b. Graph from #5

c. Graph from #7

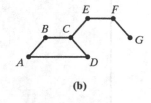

(a)

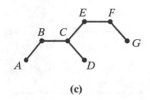

(b)

(c)

FIGURE 14.51

302

Solved Problem:

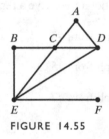

FIGURE 14.55

Find a spanning tree for the graph in **Figure 14.55** (p. 927).

A spanning tree must contain all six vertices shown in the connected graph in Figure 14.55. The spanning tree must have one edge less than it has vertices, so it must have five edges. The graph in Figure **14.55** has eight edges, so we must remove three edges. We elect to remove the edges of the circuit C, D, E, C. This leaves us the following spanning tree.

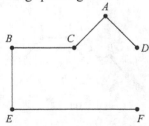

Your Turn (exercise #21, pg. 931):

Find a spanning tree for the connected graph. Be sure that your spanning tree is connected, contains no circuits, and that each edge is a bridge, with one fewer edge than vertices.

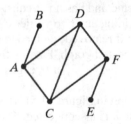

- **Objective 3 - Find the minimum spanning tree for a weighted graph.**

Solved Problem:

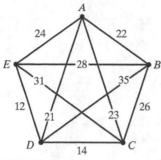

FIGURE 14.60

Use Kruskal's Algorithm to find the minimum spanning tree for the graph in **Figure 14.60** (p. 929). Give the total weight of the minimum spanning tree.

Step 1. Find the edge with the smallest weight. This is edge DE; mark it.

Step 2. Find the next-smallest edge in the graph. This is edge DC; mark it.

Step 3. Find the next-smallest edge in the graph that does not create a circuit. This is edge DA; mark it.

Step 4. Find the next-smallest edge in the graph that does not create a circuit. This is AB; mark it.

The resulting minimum spanning tree is complete. It contains all 5 vertices of the graph, and has 5 − 1, or 4, edges.

Its total weight is 12 + 14 + 21 + 22 = 69. It is shown below.

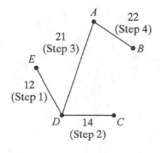

Your Turn (exercise #25, pg. 931):

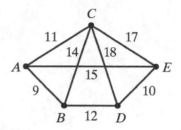

Use Kruskal's Algorithm to find the minimum spanning tree for the weighted graph. Give the total weight of the minimum spanning tree.

Chapter 1 Problem Solving and Critical Thinking

SECTION 1.1

> ### *Objective 1 - Use and understand inductive thinking.*

Check Point 1

In this video, you learn about finding counter-examples. This means
_____. Record the non-example here.

Check Point 2

a. 3, 9, 15, 21, 27….
b. 2, 10, 50, 250…..
c. 3, 6, 18, 72, 144, 432, 1728…
d. 1, 9, 17, 3, 11, 19, 5, 13, 21…

Complete the patterns for a, b, c, & d. Also summarize in words the pattern discovered for each example.

Check Point 3

As in the last check point, you are exploring patterns in lists of numbers. Complete the patterns as the video shows.

a. 1, 3, 4, 7, 11, 18, 29, 47…
b. 2, 3, 5, 9, 17, 33, 65, 129….

Now, generate a new sequence similar to each sequence but with a different starting value.
a.
b.

Check Point 4

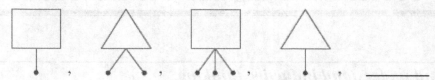

Use the video to help complete the next figure in the sequence.
Write two ways in which *inductive reasoning* can be used. Give examples as needed.

Objective 2 - Use and understand deductive reasoning.

Check Point 5
Write your conjecture about the number results here.

SECTION 1.2

Objective 1: Use estimation techniques to arrive at an approximate answer to a problem.

Check Point 1
After watching the video, use the rounding techniques learned to round the world population to the nearest
Hundred thousand:
Million:
Ten million:

Check Point 2
What is a key difference between rounding whole numbers and rounding decimals? Use an example to support your work.

Check Point 3
Think about your typical order at a fast food restaurant. Use those values to round to the nearest dollar and estimate your total bill.

Check Point 4
Assume you got a new part-time job earning $12 an hour. Use the techniques learned in the video to estimate your weekly and annual salary assuming an average of 15 hours a week.

Objective 2: Apply estimation techniques to information given by graphs.

Check Points 5 & 6
Summarize key tips regarding interpretation of graphs. Show your work for computation in check point 6.

Check Point 7
Record your answers for a, b, & c.

a.

b.

c.

Objective 3: Develop mathematical models that estimate relationships between variables.

Check Points 5 & 6
Record the mathematical model for tuition and fees. Does this seem accurate to you for colleges in your area? Why or why not?

SECTION 1.3

> ## Objective 1: Solve problems using the organization of the four-step problem-solving process.

Check Points 1 & 2

After watching the videos and thinking about the **You Try It** problems in the text, write your own problem here that is missing information or has extra information. Then explain how it should be solved, assuming that information is given.

Check Point 3

Use strategies learned in the video to determine how long it would take you to pay off the $12,850 still owed for your car if you pay $225 monthly.

Check Point 4

Show your work done in solving for the number of coin combinations for tolls.

Check Point 5

Using what you learned about combinations, how many different types of meals could a customer order if there are 5 appetizers, 8 entrees, and 4 desserts available, and he chooses one of each?

Check Point 6

The traveling salesperson problem is a typical application of graph theory (chapter 15). Sketch your solution for this salesman to travel and show his final costs for travel.

Chapter 2 Set Theory

SECTION 2.1

Objective 1: Use three methods to represent sets.

Check Points 1, 2, and 3
After watching the three check point videos and reviewing the material, use all three methods to represent the set A defined as {0, 2, 4, 6, 8, 10…..}

Description of the Set:

Roster:

Set Notation:

Objective 2: Define and recognize the empty set.

Check Point 4
Which of the answers is the empty set? Explain your answer as learned in the video.

Objective 3: Use the symbols ∈ and ∉.

Check Point 5

Use the video and your text to determine when each symbol is appropriate, and summarize that information here.

Objective 4: Apply set notation to sets of natural numbers.

Check Point 6

Record the answer in set notation for each of the examples.

a.

b.

c.

Check Point 7

Use the information given to record the appropriate notation for each set.

a.

b.

Objective 5: Determine a set's cardinal number.

Check Point 8

After learning about determining cardinal number of the sets, summarize how one should determine the cardinality of a set.

Objective 6: Recognize equivalent sets.

Check Point 9
How do we know if sets are equivalent?

Objective 7: Distinguish between finite and infinite sets.

No check point video is given here. Use this space to note how we can distinguish between finite and infinite sets.

Objective 8: Recognize equal sets.

Check Point 10
How do we know if two sets are equal? What is a tip shared in the video?

SECTION 2.2

Objective 1: Recognize subsets and use the notation ⊆.

Check Point 1

Record your answers for a, b, & c. Use the video and your text to determine how you can know whether a set is a subset or not.

a.

b.

c.

Objective 2: Recognize proper subsets and use the notation ⊂.

Check Point 2

Record the correct answers for parts a & b and write an explanation of the difference between a subset and proper subset.

a.

b.

Check Point 3

What did you learn about the empty set as a subset?

Objective 3: Determine the number of subsets of a set.

Check Point 4
Record your answers for a & b and summarize how to determine the number of subsets and proper subsets.

a.

b.

Objective 4: Apply concepts of subsets and equivalent sets to infinite sets.

There is no check point video here. However, use you what you learned in your text and the check point videos to determine the number of subsets for infinite sets and determining equivalent sets for infinite sets. Explain your reasoning.

SECTION 2.3

> ## Objective 1: Understand the meaning of a universal set.

There is no check point video here. What is the universal set? Describe it in your own words.

> ## Objective 2: Understand the basic ideas of a Venn diagram.

Check Point 1
Record your answers to the questions here.
a.   U =

b.   A=

c.   Elements in U not in A =

> ## Objective 3: Use Venn diagrams to visualize relationships between two sets.

Check Point 2
Record your answers here.
a.   A

b.   The set of elements in B but not in A

c.   The set of elements in U that are not in A

d.   The set of elements in U that are not in A or B

Objective 4: Find the complement of a set.

Check Point 3

What is the complement of A here? What does complement mean?

A´ =

Objective 5: Find the intersection of two sets.

Check Point 4

Record the answers here to intersections of the sets.

a.

b.

c.

Can you summarize in words how to find an intersection of sets?

Objective 6: Find the union of two sets.

Check Point 5

Record your answers here to finding unions of sets.

a.

b.

c.

How do you find a union of sets?

Objective 7: Perform operations with sets.

Check Point 6
Find (A U B)′

Find A′∩ B′

Do you see any patterns here?

Objective 8: Determine sets involving set operations from a Venn diagram.

Check Point 7
Use the check point video and the diagram to find the results from set operations.

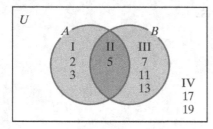

a. $A \cap B$

b. $(A \cap B)'$

c. $A \cup B$

d. $(A \cup B)'$

e. $A' \cup B$

f. $A \cap B'$.

a.

b.

c.

d.

e.

f.

Objective 9: Understand the meaning of "and" and "or".

There is no check point video here. Why is it important to be precise about the use of English language in working in set problems?

Objective 10: Use the formula for n (A U B).

Check Point 8
Use the video to find the number of students who were registered in a math class or an English class. Sketch the diagram and show work.

SECTION 2.4

Objective 1: Perform set operations with three sets.

Check Point 1
Use the videos to record the results for the following set operations.
a. $A \cup (B \cap C)$

b. $(A \cup B) \cap (A \cup C)$

c. $A \cap (B \cup C')$

Objective 2: Use Venn diagrams with three sets.

Check Point 2
Use the diagram and the video to record the results for a-e. Show work as needed.

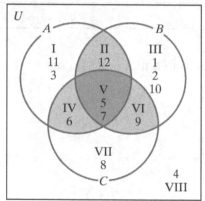

a. C
b. B ∪ C
c. A ∩ C
d. B′
e. A ∪ B ∪ C

Check Point 3
Use your video to construct a Venn diagram for the following sets.
A = {1, 3, 6, 10}
B = {4, 7, 9, 10 }
C = (3,4,5,8, 9, 10}
U = {1,2,3,4,5,6,7,8, 9, 10}

Objective 3: Use Venn diagrams to prove equality of sets.

In Check Point 4, you learn about equality of sets. What did you conclude?

In Check Point 5, what did you conclude about the sets? Sketch the diagrams from the video.

SECTION 2.5

Objective 1: Use Venn diagrams to visualize a survey's results.

Check Point 1

Use the video and the diagram to answer the questions here.

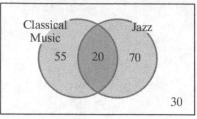

Objective 2: Use survey results to complete Venn diagrams and answer questions about the survey.

Check Point 2

Use the video and diagram to determine how many men agreed with the question.

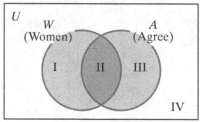

Check Points 3 & 4

Draw the Venn diagram used in the video to display the results from the survey of memorabilia collectors. Be sure to label components accurately. Then use the diagram to find the answers to the questions.

How many people collect:

a. Comic books, but neither baseball cards nor stamps

b. Baseball cards and stamps, but not comic books

c. Baseball cards or stamps, but not comic books

d. Exactly two types of memorabilia

e. At least one type of memorabilia

f. None of the types of memorabilia.

Chapter 3 Logic

SECTION 3.1

Objective 1: Statements, Negations, and Quantified Statements

There is no check point video for this objective. What is a **statement**?

Objective 2: Express statements using symbols

There is no check point video for this objective. Write original statements for 'p' and 'q.'

p:

q:

Objective 3: Form the negation of a statement

Check Point 1
Use your check point video to write the negations of the two statements here.
a.

b.

Objective 4: Express negations using symbols
Check Point 2
Use your check point video to write the statements symbolically.
a.

b.

Objective 5: Translate a negation represented by symbols into English

Check Point 3
Use your check point video to express ~ q in words below.

~ q :

Objective 6: Express quantified statements in two ways

There is no check point video for this section. Use the figure to summarize how to write quantified statements.

Objective 7: Write negations of quantified statements.

Check Point 4
In this video, you learn how to express the statement in two ways. Write them here. What do you need to remember about negating quantified statements?

SECTION 3.2

Objective 1: Express compound statements in symbolic form

Check Point 1
Write the compound statements in symbolic form.

a.

b.

Check Point 2
Write the compound statements in symbolic form as shown in your video.

a.

b.

Check Point 3
Use your videos to write the if/then statements in symbolic form.

a.

b.

Check Point 4
Use your video to write the compount statement in symbolic form. "Suffering huge budget deficits is necessary for not controlling military spending."

Check Point 5
Use your video and the following statements to write the compounds statements in symbolic form.   p: The word is *run*.   q: the word has 396 meanings.

a.

b.

Objective 2: *Express symbolic statements with parentheses in English*

Check Point 6

Use your video to help you write the following statements in words.

a. $\sim(p \wedge q)$

b. $\sim(q \wedge p)$

c. $\sim(q \rightarrow p)$

Check Point 7

Use your video to write the statements in words.

a. $(p \wedge \sim q) \rightarrow \sim r$

b. $p \wedge (\sim q \rightarrow \sim r)$

Objective 3: *Use the dominance of connectives.*

Check Point 8

Use your video to help you write the compound statements in symbolic form.

a.   If there is too much homework or a teacher is boring, then I do not take that class.

b.   There is too much homework, or if a teacher is boring, then I do not take that class.

SECTION 3.3

Objective 1: Use the definition of negation, conjunction, and disjunction.

Check Point 1
Use your video to determine the truth value for each of the following.

a. $p \wedge q$

b. $p \wedge \sim q$

c. $\sim p \vee q$

d. $\sim p \vee \sim q$

Objective 2: Construct truth tables.

Check Point 2
Use your video to construct a truth table for $\sim (p \vee q)$ to determine when the statement is true and when the statement is false.

Check Point 3
Record the truth table for $\sim p \wedge \sim q$ here.

Check Point 4
Record the truth table here. Write two tips to keep in mind when completing truth tables.

Check Point 5
What happens when you combine p and $\sim p$?

Check Point 6
Record the answers to the questions here.

a.

b.

Objective 3: Determine the truth value of a compound statement for a specific cause

Check Point 7
Use the information in the circle graph in the text and in the video to determine the truth value for the statement. Record the result here.

SECTION 3.4

Objective 1: Understand the logic behind the definition of the conditional

There is no check point video for this objective. When is a conditional statement false?

Objective 2: Construct truth tables for conditional statements

Check Point 1

Show the truth table for $\sim p \sim q$ using the check point video.

Check Point 2

Construct the truth table as shown in the video. How do we know a statement is a tautology?

Check Point 3

Is the claim in the advertisement false? How do we know?

Objective 3: Understand the definition of the biconditional

There is no check point video for this objective. When is a biconditional TRUE?

Objective 4: Construct truth tables for biconditional statements
---

Check Point 4
Construct the truth table below, as shown in the video. Is it a tautology?

Objective 5: Determine the truth value of a compound statement for a specific case
---

Check Point 5
What is the claim's truth value?

SECTION 3.5

Objective 1: Use a truth table to show that statements are equivalent

---

Check Point 1

Use your video to complete the following.

a. Show that $p \lor q$ and $\sim q \rightarrow p$ are equivalent

b. Use that result to write a statement equivalent to "I attend classes or I lose my scholarship."

Check Point 2

Use your video to show that $\sim [\sim (\sim p)] \equiv \sim p$

Check Point 3

Which statement is NOT equivalent to "If it's raining, then I need my jacket?"

Objective 2: Write the contrapositive for a conditional statement

Check Point 4
Use your video to record the contrapositives for each of the statements.
a.

b.

c.

d.

Objective 3: Write the converse and inverse of a conditional statement

Check Point 5
Write the converse, inverse, and contrapositive of the statement.
"If you are in Iran, then you don't see a Club Med."

SECTION 3.6

Objective 1: Write the negation of a conditional statement

Check Point 1
Use your video to write the negation of "If you do not have a fever, you do not have the flu."

Objective 2: Use DeMorgan's Law

Check Point 2
Write the equivalent statement here.

Check Point 3
Write the equivalent statement here.

Check Point 4
Write the negation for the two statements.
a)

b)

Check Point 5
Use the video to write the equivalent statement to "If it is not windy, we can swim and we cannot sail."

SECTION 3.7

Objective 1: Use truth tables to determine validity

Check Point 1
Use the video to construct the truth table for the argument. What is the conclusion?

Check Point 2
Use the video to complete the truth table. Is the argument valid or invalid?

I study for 5 hours or I fail.
I did not study for 5 hours.
Therefore, I failed.

Check Point 3
Is the argument valid or invalid?

Objective 2: Recognize and use forms of valid and invalid arguments
---

Check Point 4
Use the video to determine whether each argument is valid or invalid. Write the reason why with each conclusion.

a.

b.

c.

Check Point 5
Use the video to determine whether the argument is valid or not. Explain why.

SECTION 3.8

Objective 1: Use Euler diagrams to determine validity

Check Points 1-6

Use the video to sketch the diagram for each check point and label as valid or invalid. Write any additional notes as needed.

1.

2.

3.

4.

5.

6.

Chapter 4 Number Representation and Calculation

SECTION 4.1

Objective 1: Evaluate an exponential expression.

Check Point 1
Use the video to evaluate the following.
a. 7^2
b. 5^3
c. 1^4
d. 10^5
e. 10^6
f. 18^1

Objective 2: Write a Hindu-Arabic numeral in expanded form.

Check Point 2
Use the video to write the follow numbers in expanded form.
a. 4026

b. 24,232

Objective 3: Express a number's expanded form as a Hindu-Arabic numeral.

Check Point 3
Express the expanded form as a Hindu-Arabic numeral
a.

b.

Objective 4: Understand and use the Babylonian numeration system.

Check Point 4

Use the video to write the numerals as Hindu-Arabic numerals. Show your calculation work.

a.

b.

Objective 5: Understand and use the Mayan numeration system.

Check Point 5

Write the two Mayan numerals as Hindu-Arabic numerals. Show work.

a.

b.

SECTION 4.2

Objective 1: Change numerals in bases other than ten to base ten.

Check Point 1

Use the video to convert 3422_{five} to base ten. Show the work.

Check Point 2

Use the video to convert 110011_{two} to base ten. Show the work.

Check Point 3

Use the video to convert $\text{AD4}_{\text{sixteen}}$ to base ten. Show the work.

Objective 2: Change base ten numerals to numerals in other bases.

Check Point 4

Use the video to convert base ten numeral 6 to base five. Show the work.

Check Point 5

Use the video to convert the base ten numeral 365 to base seven. Show the work.

Check Point 6

Use the video to convert the base ten numeral 51 to base two. Show the work.

Check Point 7

Use the video to convert the base ten numeral 2,763 to base five. Show the work.

SECTION 4.3

Objective 1: Add in bases other than ten.

Check Point 1

Use the video to help you add the two numbers in base five. Show your work.

Check Point 2

Use the video to help you add the two numbers in base two. Show your work.

Objective 2: Subtract in bases other than ten.

Check Point 3

Use the video to help you subtract the two numbers in base five. Show your work.

Check Point 4

Use the video to help you complete the subtraction in base seven. Show your work. Are there any tips to keep in mind when adding/subtracting in other bases?

Objective 3: Multiply in bases other than ten.

Check Point 3

Use the video to help you subtract the two numbers in base five. Show your work.

Check Point 4

Use the video to help you complete the subtraction in base seven. Show your work. Are there any tips to keep in mind when adding/subtracting in other bases?

Objective 4: Understand and use the Ionic Greek system.

Check Point 7

Use the video to write $\omega\pi\epsilon$ as a Hindu-Arabic numeral.

Chapter 5 Number Representation and Calculation

SECTION 5.1

Objective 1: Determine divisibility.

Check Point 1
Which of the statements are true? Explain why/why not.

a)

b)

c)

Objective 2: Write the prime factorization of a composite number.

Check Point 2
Find the prime factorization of 120.

Objective 3: Find the greatest common divisor of two numbers.

Check Point 3
Use the video to help you find the greatest common factor of 225 and 825.

Objective 4: Solve problems using the the greatest common divisor.

Check Point 4

What is the largest number of people in the singing group?

Objective 5: Find the least common multiple of two numbers.

Check Point 5

Use the video to answer the question. Find the least common multiple (LCM) of 18 and 30.

Objective 6: Solve problems using the least common multiple.

Check Point 6

Use the video to find out when the movies will begin at the same time.

SECTION 5.2

Objective 1: Define the set of integers.

There is no check point video for this objective. Describe the set of integers in your own words.

Objective 2: Graph integers on a number line.

Check Point 1
Sketch the number line and plot the points as shown in the video.

Objective 3: Use the symbols < and >.

Check Point 2
Insert < or > to make the statement true.
a.   6_____ −7
b.   −8_____ −1
c. −25_____ −2
d. −14_____ 0

Objective 4: Find the absolute value of an integer.

Check Point 3
Use the video to find the absolute value. Write a tip about working with absolute values.
a. $|-8|$
b. $|6|$
c. $-|8|$

Objective 5: *Perform operations with integers.*

Check Point 4
Write the solutions to the problems and a tip about addition and subtraction of integers.
a.

b.

c.

Check Point 5
Use the video and figure to answer the question. Record your work here.

Check Point 6
Evaluate the exponential expressions. What can you conclude about exponents and negatives?
a. $(-5)^2$

b. -5^2

c. $(-4)^3$

d. $(-3)^4$

Objective 6: *Use the order of operations agreement.*

Check Points 7 & 8
Use the videos and order of operations to simplify the expression.

$$7^2 - 48 \div 4^2 \times 5 + 2$$

$$(-8)^2 - (10 - 13)^2 (-2)$$

SECTION 5.3

Objective 1: Define rational numbers.

There is no check point video for this objective. Write a definition and examples of rational numbers.

Objective 2: Reduce rational numbers.

Check Point 1

Reduce $\dfrac{72}{90}$ to lowest terms.

Objective 3: Convert between mixed numbers and improper fractions.

Check Point 2

Convert the mixed number $2\dfrac{5}{8}$ to an improper fraction using the video lesson. What is an important tip to keep in mind?

Check Point 3

Convert $\dfrac{5}{3}$ to a mixed number.

Objective 4: Express rational numbers as decimals.

Check Point 4

Express the fractions $\frac{3}{8}$ and $\frac{5}{11}$ as decimals.

Objective 5: Express decimals in the form a/b.

Check Point 5

Express the terminating decimals as fractions in lowest terms.

a.

b.

c.

Check Points 6 & 7

Express the repeating decimal 0.22…… as a quotient of integers. Show work.

Express the repeating decimal 0.797979…..as a quotient of integers. Show work.

Objective 6: Multiply and divide rational numbers.

Check Point 8

Use the video to multiply the fractions and reduce to lowest terms.

a.

b.

c.

Check Point 9
Use the video to help you divide and reduce the fractions.

a.

b.

c.

Objective 7: Add and subtract rational numbers.

Check Point 10
Add and reduce your fractions.

a.

b.

c.

Check Point 11
Add the fractions with unlike denominators. Show work.

$$\frac{3}{10} + \frac{1}{6}$$

Check Point 12
Subtract the fractions and show work.

$$\frac{3}{10} - \frac{7}{12}$$

Objective 8: Use the order of operations agreement with rational numbers.

Check Point 13
Use the video to help you with the problem. Show work.

$$\left(-\frac{1}{2}\right)^2 - \left(\frac{7}{10} - \frac{8}{15}\right)^2 (-18)$$

Objective 9: Apply the density property of rational numbers.

Check Point 14

Find the rational number halfway between $\frac{1}{3}$ and $\frac{1}{2}$.

Objective 10: Solve problems involving rational numbers.

Check Point 15
How many eggs do you need for the recipe?

SECTION 5.4

Objective 1: Define irrational numbers.

There is no check point video here. Write a definition of irrational numbers in your own terms.

Objective 2: Simplify square roots.

Check Point 1
Simplify the square roots as shown in the video.

$\sqrt{12}$ $\qquad$ $\sqrt{60}$ $\qquad$ $\sqrt{55}$

Objective 3: Perform operations with square roots.

Check Point 2
Multiply and simplify as shown in the video.

a. $\sqrt{3} \cdot \sqrt{10}$ $\qquad$ b. $10 \cdot \sqrt{10}$ $\qquad$ c. $\sqrt{6} \cdot \sqrt{2}$

Check Point 3
Find the quotients as shown in the video.

a. $\dfrac{\sqrt{80}}{\sqrt{5}}$ $\qquad$ b. $\dfrac{\sqrt{48}}{\sqrt{6}}$

Check Points 4 & 5

Add or subtract as shown in the video.

a. $8\sqrt{3} + 10\sqrt{3}$

b. $4\sqrt{13} - 9\sqrt{13}$

c. $7\sqrt{10} + 2\sqrt{10} - \sqrt{10}$

a. $\sqrt{3} + \sqrt{12}$ b. $4\sqrt{8} - 7\sqrt{18}$

Objective 4: Rationalize denominators.

Check Point 6

Rationalize the denominators as shown in the video.

a. $\dfrac{25}{\sqrt{10}}$ b. $\dfrac{2}{\sqrt{7}}$ c. $\dfrac{5}{\sqrt{18}}$

SECTION 5.5

Objective 1: Recognize subsets of real numbers.

Check Point 1
Use the video to help you classify the numbers in the set as given.

Consider the following set of numbers:
$$\left\{-9, -1.3, 0, 0.\overline{3}, \frac{\pi}{2}, \sqrt{9}, \sqrt{10}\right\}.$$

List the numbers in the set that are
    a. natural numbesl   **b.** whole numbers.
    c. integers.   **d.** rational numbers.
    e. irrational numbers.   **f.** real numbers.

Objective 2 Recognize properties of real numbers.

Check Point 2
Use the video to help you name the properties illustrated.

    a. $(4 \cdot 7) \cdot 3 = 4 \cdot (7 \cdot 3)$

    b. $3\left(\sqrt{5} + 4\right) = 3\left(4 + \sqrt{5}\right)$

    c. $3\left(\sqrt{5} + 4\right) = 3\sqrt{5} + 12$

    d. $2\left(\sqrt{3} + \sqrt{7}\right) = \left(\sqrt{3} + \sqrt{7}\right)2$

    e. $1 + 0 = 1$

    f. $-4\left(-\dfrac{1}{4}\right) = 1.$

Check Point 3
Use the video to answer the following. Explain your answers.
a. Are the natural numbers closed with respect to multiplication?

b. Are the integers closed with respect to division?

Check Point 4

Answer the questions below.

a. How can you tell the set is closed under addition?

b. Verify the associative property.

c. What is the identity element in the 4 hour clock system?

d. Find the inverse of each element in the 4-hour clock system.

e. Verify two cases of the commutative property as shown.

SECTION 5.6

Objective 1: Use properties of exponents.

Check Point 1
Use the video to help you simplify the following. Summarize what you learned about the use of zero as an exponent.
a. 19^0

b. $(3\pi)^0$

c. $(-14)^0$

d. -14^0

Check Point 2
Use the videos to help you simplify the following.
a. 9^{-2}

b. 6^{-3}

c. 12^{-1}

Objective 2: Convert from scientific notation to decimal notation.

Check Point 3
Use the videos to convert the following.
a. 7.4×10^9
b. 3.017×10^{-6}

Objective 3: Convert from decimal notation to scientific notation.

Check Point 4
Use the video to help convert to scientific notation.

   a. 7,410,000,000
   b. 0.000000092

Check Point 5
As of December 2008, the US population was 308 million. Write that in scientific notation as shown in the video.

Objective 4: Perform computations using scientific notation.

Check Point 6
Multiply and write the answer in decimal notation as shown in the video.
$$\left(1.3\times10^{7}\right)\left(4\times10^{-2}\right)$$

Check Point 7
Divide and write the answer in decimal notation as shown.
$$\frac{6.9\times10^{-8}}{3\times10^{-2}}$$

Check Point 8
Write the answer in both scientific and decimal notation. 0.0036 x 5,200,000.

Objective 5: Solve applied problems using scientific notation.

Check Point 9
Use the video to answer how much each citizen would have to pay

SECTION 5.7

| **Objective 1:** Write terms of an arithmetic sequence. |

Check Point 1
Use the video to help you write the first six terms of the sequence with first term of 100 and common difference of 20.

Check Point 2
Write the first six terms of the sequence where $a_1 = 8$ and $d = -3$.

| **Objective 2:** Use the formula for the general term of an arithmetic sequence. |

Check Point 3
Find the ninth term of the sequence where the first term is 6 and the difference is -5.

Check Point 4
Use the bar graph and video to write a formula for the nth term and project the percentage of full-time tenured college faculty for 2020.

Objective 3: Write terms of a geometric sequence.

Check Point 5
Write the first six terms of the geometric sequence with the first term 12 and the common ratio of $\frac{-1}{2}$.

Objective 4: Use the formula for the general term of a geometric sequence.

Check Point 6
Find the seventh term of the geometric sequence whose first term is 5 and the common ratio is -3.

Chapter 6 Algebra Equations and Inequalities

SECTION 6.1

Objective 1: Evaluate algebraic expressions.

Check Points 1, 2, & 3: Evaluate as shown in the videos.

1. $8 + 6(x - 3)^2$ for $x = 13$
2. $x^2 + 4x - 7$ for $x = -5$
3. $-3x^2 + 4xy - y^3$ for $x = 5$ and $y = -1$

Objective 2: Use mathematical models.

Check Point 4

The mathematical model $M = -120x^2 + 998x + 590$ describes the calories burned per day, M, by a man in age group x with moderately active lifestyles. Use the video to find out how many calories per day are needed. How does this compare to the graph in Figure 6.1?

Objective 3: Understand the vocabulary of algebraic expressions.

There are no check point videos for this objective. Use this area to jot notes about terms such as:

- Terms
- Coefficient
- Constant
- Factors
- Like terms

Objective 4: Simplify algebraic expressions..

Check Point 5
Use the video to simplify $7(2x - 3) - 11x$.

Check Point 6
Simplify $7(4x^2 + 3x) + 2(5x^2 + x)$

Check Point 7
Simplify $6x + 4[7 - (x - 2)]$

SECTION 6.2

Objective 1: Solve linear equations.

Check Point 1
Solve and check using the video. $4x + 5 = 29$

Check Point 2
Solve and check using the video. $6(x - 3) - 10x = -10$

Check Point 3
Solve and check using the video. $2x + 9 = 8x - 3$

Check Point 4
Solve and check using the video. $4(2x + 1) = 29 + 3(2x - 5)$

Objective 2: Solve linear equations containing fractions.

Check Point 5

Solve and check using the video. $\dfrac{2x}{3} = 7 - \dfrac{x}{2}$

Check Point 6

Use the video to find out the intensity of the event for a level of depression of 10 for the low-humor group. How is the solution shown on the blue line graph in the figure?

Objective 3: Solve proportions.

Check Point 7

a. $\dfrac{10}{x} = \dfrac{2}{3}$

b. $\dfrac{22}{60-x} = \dfrac{2}{x}$

Objective 4: Solve problems using proportions.

Check Point 8

Use the video to determine the property tax on a house with assessed value of $420,000.

Check Point 9
Use the video to help develop the correct proportion and find the number of deer in the refuge.

Objective 5: *Identify equations with no solution or infinitely many solutions.*

Check Point 10
Solve $3x + 7 = 3(x + 1)$ and explain the results using the video.

Check Point 11
Solve and check using the video. What do the results mean?
$7x + 9 = 9(x + 1) - 2x$

SECTION 6.3

Objective 1: Use linear equations to solve problems.

Check Point 1

Use the bar graph and video to find the average yearly salary of women with the levels of education noted. Show work.

Check Point 2

Use the information in the figure and on the video to find by what year only 33% of female freshmen consider the meaningful philosophy essential or very important.

Check Point 3

Use the equations to find the number of minutes at which long-distance calls for the two plans are the same cost.

Check Point 4

Use the equations to find the cost of the computer before the reduction in price.

Objective 2: Solve a formula for a variable.

Check Point 5

Solve the $P = 2l + 2w$ for w. Show work.

Check Point 6

Solve $T = D + pm$ for m. Show work.

SECTION 6.4

Objective 1: Graph subsets of real numbers on a number line.

Check Point 1

Use the video to help you show the solutions for each inequality on a number line.

a. $x < 4$

b. $x \geq -2$

c. $-4 \leq x \leq 1$

Objective 2: Solve linear inequalities.

Check Point 2

Solve and graph, using the video. $5x - 2 \leq 17$.

Check Point 3

Solve and graph the inequalities. What do you need to remember about showing the solutions on the number line?

Check Points 4, 5, & 6
Solve and graph the solutions, using the videos.
$7x - 3 > 13x + 33$

$2(x - 3) - 1 \leq 3(x + 2) - 14$

$1 \leq 2x + 3 \leq 11$

Objective 3: Solve applied problems using linear inequalities.

Check Point 7
What grade must you earn on the final to get a B in the course? Show work from the video.

Section 6.5

Objective 1: Multiply binomials using the FOIL method.

Check Point 1

Multiply $(x + 5)(x + 6)$

Check Point 2

Multiply $(7x + 5)(4x - 3)$

Objective 2: Factor trinomials.

Check Point 3

Factor $x^2 + 5x + 6$

Check Point 4

Factor $x^2 + 3x - 10$

Check Point 5

Factor $5x^2 - 14x + 8$

Check Point 6

Factor $6y^2 + 19y - 7$

Objective 3: Solve quadratic equations by factoring.

Check Point 7

Solve $(x + 6)(x - 3) = 0$

Check Point 8

Solve $x^2 - 6x = 16$

Check Point 9

Solve $2x^2 + 7x - 4 = 0$

Objective 4: Solve quadratic equations using the quadratic formula.

Check Point 10

Solve using the quadratic formula. $8x^2 + 2x - 1 = 0$

Check Point 11

Solve using the quadratic formula. $2x^2 = 6x - 1$.

Objective 5: Solve problems modeled by quadratic equations.

Check Point 12

Find the age of the woman with a blood pressure of 115 mm Hg.

Chapter 7 Algebra Graphs, Functions and Linear Systems

SECTION 7.1

Objective 1: Plot points in the rectangular coordinate system.

Check Point 1
Use the video to plot points as shown.

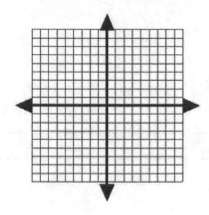

http://algebranotes.info/printable-graph-paper/ is a great place to print graph paper if you need some!

Objective 2: Graph equations in the rectangular coordinate system.

Check Point 2
Graph $y = 4 - x$ as shown in the video.

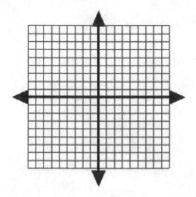

Check Point 3
Make a table of values as shown in the video, then graph the equations. What are the coordinates of the intersection points? What does this mean, practically speaking?

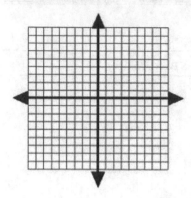

Objective 3: Use function notation.

Check Point 4
Use the video to complete the following.

a. $f(6)$ for $f(x) = 4x + 5$
b. $g(-5)$ for $g(x) = 3x^2 - 10$
c. $h(-4)$ for $h(r) = r^2 - 7r + 2$

Check Point 5
Use the line graph and video to

a. estimate the car's required stopping distance at 40 mph on dry pavement.

b. Use the function $f(x) = 0.0875x^2 - 0.4x + 66.6$ to find the required stopping distance at 40 mph

Objective 4: Graph functions.

Check Point 6

Use the video to graph the functions $f(x) = 2x$ and $g(x) = 2x - 3$ in the same system as shown. How do the graphs compare?

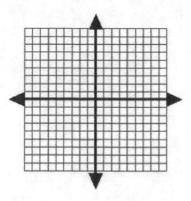

Objective 5: Use the vertical line test.

Check Point 7

Use your video to identify the graphs that are functions. What is a rule of thumb to use?

Use the vertical line test to identify graphs in which y is a function of x.

a.

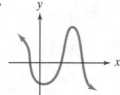

b.

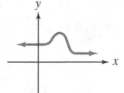

c.

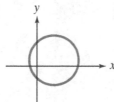

Objective 6: *Obtain information about a function from its graph.*

Check Point 8

Use the video and figure shown to answer the questions.

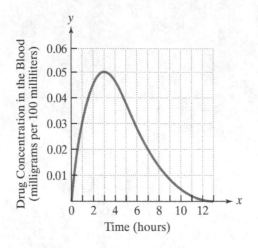

a.   During which period of time is the drug concentration increasing?
b.   During which period of time is it decreasing?
c.   What is the drug's maximum concentration? When does this occur?
d.   What happens by the end of 13 hours?
e.   Why is this graph defining a function?

SECTION 7.2

Objective 1: Use intercepts to graph a linear equation.

Check Point 1

Graph $2x + 3y = 6$ as shown in the video, using intercepts.

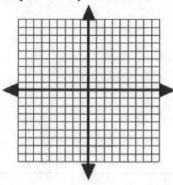

Objective 2: Calculate slope.

Check Point 2

Calculate slope for the pairs of points as shown.

a. $(-3, 4)$ and $(-4, -2)$
b. $(4, -2)$ and $(-1, 5)$

Objective 3: Use the slope and y-intercept to graph a line.

Check Point 3

Graph as shown using slope and y-intercept. $y = \dfrac{3}{5}x + 1$

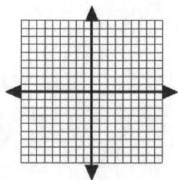

Check Point 4

Graph $3i + 4y = 0$ using the slope and y-intercept, as shown.

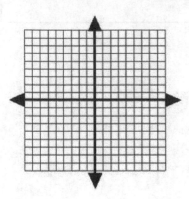

Objective 4:   Graph horizontal or vertical lines.

Check Point 5

Graph $y = 3$.

Check Point 6

Graph $x = 2$.

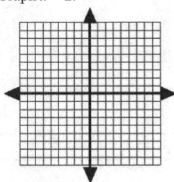

Objective 5: Interpret slope as rate of change.

Check Point 7
Find the slope of the line. What does it mean?

Objective 6: Use slope and y-intercept to model data.

Check Point 8
Use the two points for college to find a function in the form $C(x) = mx + b$ to model the percentage of college graduates in the U.S. population in x years after 1960. Then use the model to project the percentage of college graduates in 2020.

SECTION 7.3

Objective 1: Decide whether an ordered pair is a solution of linear system.

Check Point 1

Is the point a solution to the system? Why or why not? How do you know if a point is the solution?

Objective 2: Solve linear systems by graphing.

Check Point 2

Solve by graphing as shown in the video.
$$\begin{cases} 2x + 3y = 6 \\ 2x + y = -2 \end{cases}$$

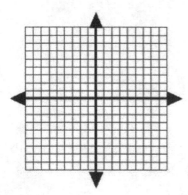

Objective 3: Solve linear systems by substitution.

Check Points 3 & 4

Solve by the substitution method as shown.

$$\begin{cases} 5x - 4y = 9 \\ x - 2y = -3 \end{cases} \qquad \begin{cases} 3x + 2y = -1 \\ x - y = 3 \end{cases}$$

Objective 4: Solve linear systems by addition.

Check Points 5 & 6
Solve as shown in the videos. Show work.

$$\begin{cases} 4x + 5y = 3 \\ 2x - 3y = 7 \end{cases} \qquad \begin{cases} 3x = 2 - 4y \\ 5y = -1 - 2x \end{cases}$$

Objective 5: Identify systems that do not have exactly one ordered-pair solution.

Check Points 7 & 8
Solve the systems. Explain the results.

$$\begin{cases} x + 2y = 4 \\ 3x + 6y = 13 \end{cases} \qquad \begin{cases} y = 4x - 4 \\ 8x - 2y = 8 \end{cases}$$

SECTION 7.4

Objective 1: *Graph a linear inequality in two variables.*

Check Point 1
Graph the inequality as shown. $2x - 4y \geq 8$

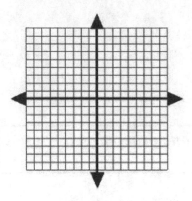

Check Point 2
Graph $y > \dfrac{-3}{4}x$.

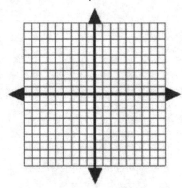

Check Point 3
Graph $y > 1$ and $x - 2$.

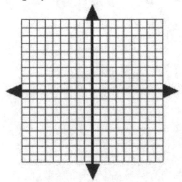

Objective 2: Use mathematical models involving linear inequalities.

Check Point 4
How do we know that point B is a solution?

Objective 3: Graph a system of linear inequalities.

Check Point 5
Show the solution.

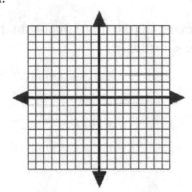

$$\begin{cases} x + 2y > 4 \\ 2x - 3y \le -6. \end{cases}$$

Check Point 6

Graph the solution set of $\begin{cases} x < 3 \\ y \ge -1 \end{cases}$

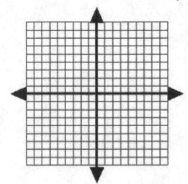

SECTION 7.5

> ### *Objective 1: Write an objective function describing a quantity that must be maximized or minimized.*

Check Point 1

Write the objective function that describes the company's total daily profit, z, from x bookshelves and y desks.

> ### *Objective 2: Use inequalities to describe limitations in a situation.*

Check Point 2

To maintain quality, the company should not manufacture more than a total of 80 desks and shelves per day. Write the inequality that represents this.

Check Point 3

Write the inequalities that describe the constraints. Then summarize what you have described about this company.

Objective 3: Use linear programming to solve problems.

Check Point 4

How many bookshelves and how many desks should be manufactured per day to obtain a maximum profit? What is the maximum daily profit? Show work.

SECTION 7.6

Objective 1: Graph exponential functions.

Check Point 1

Graph $f(x) = 3^x$

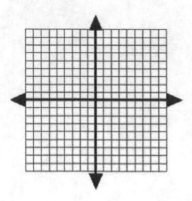

Objective 2: Use exponential models.

Check Point 2

How well did the functions model world population? What was a better model for the prediction?

Check Point 3

What is the risk of a car accident with a blood alcohol concentration of 0.01? Round to one decimal place.

Objective 3: Graph logarithmic functions.

Check Point 4

Rewrite the logarithmic function in exponential form as shown. Then obtain the graph.

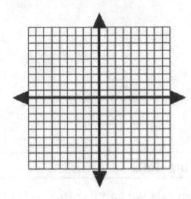

Objective 4: Use logarithmic models.

Check Point 5

How well does the function model the actual increase shown in the figure?

Objective 5: Graph quadratic functions.

Check Point 6

Graph $y = x^2 + 6x + 5$

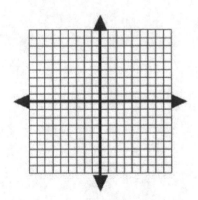

Objective 6: Use quadratic models.

Check Point 7
How far would he need to reach to block the punt?

Objective 7: **Determine an appropriate function for modeling data.**

There is no check point video for this section. Describe some methods we can use to determine an appropriate function for modeling data. What are some considerations to keep in mind?

Chapter 8 Personal Finance

SECTION 8.1

Objective 1: *Express a fraction as a percent.*

Check Point 1
Express 1/8 as a percent.

Objective 2: *Express a decimal as a percent.*

Check Point 2
Express 0.023 as a percent.

Objective 3: *Express a percent as a decimal.*

Check Point 3
Express the percents as decimals.
a.   67%

b.   250%

Objective 4: Solve applied problems involving sales tax and discounts.

Check Point 4
Suppose the tax rate is 6% and you purchase a computer for $1260. Use the video to help you find the tax and total cost paid. Show work.

Check Point 5
Use the video to find how the discount amount and sales price for the CD player. Show work.

Objective 5: Compute income tax.

Check Point 6
Use the 2008 marginal tax rates table and video to find the tax owed by the single man whose financial information is given. Show work.

Objective 6: Determine percent increase or decrease.

Check Point 7

Use the video to find the percentage increase or decrease.

a.   If 6 is increased to 10.

b.   If 10 is decreased to 6.

Check Point 8

Use the video to help solve the problem. A television that regularly sells for $940 is on sale for $611. What is the percent decrease?

Objective 7: Investigate some of the ways percent can be abused.

Check Point 9

An episode of a TV series had an audience of 12% versus its usual 10%. What is the percent increase for this episode? Use the video to complete the problem.

Check Point 10

Use the video to answer what taxes are paid in year 2 and what percentage increase or decrease there is. Show work.

SECTION 8.2

Objective 1: Determine gross income, adjustable income, and taxable income.

Check Point 1

a.   Determine the woman's gross income.

b.   Determine the woman's adjusted gross income.

c.   Determine the woman's taxable income.

Objective 2: Calculate federal incomes tax.

Check Point 2

Calculate the tax owed by the man as shown.

Objective 3: Calculate FICA taxes.

Check Point 3

If you are not self-employed and earn $200,000, what will be your FICA taxes?

Objective 4: Solve problems involving working students and taxes.

Check Point 4
Use the video to find the following.
a. The weekly gross pay.

b. The amount of federal tax withheld.

c. FICA taxes withheld.

d. State taxes withheld

e. Weekly net pay

f. Overall percentage of pay withheld for taxes?

SECTION 8.3

Objective 1: Calculate simple interest.

Check Point 1

Use the video to find the interest on $3000 at 5% at the end of the first year. Show work.

Check Point 2

Use the video to find the interest on the loan of $2400 for 2 years at 7%.

Objective 2: Use the future value formula.

Check Point 3

Find the loan's future value, as shown in the video. Show work. The loan of $2040 was made at 7.5% for four months.

Check Point 4

What simple interest rate would you pay if you borrowed $5000 and paid back $6800 as promised? Show the work.

Check Point 5

How much should you put in an investiment paying a simple interest rate of 8% if you needed $4000 in six months? Use the video to solve, and show your work.

SECTION 8.4

Objective 1: Use compound interest formulas.

Check Point 1

You deposit $1000 in a savings account that has a rate of 4%. Use the video to find the amount A of money after 5 years if the interest is compounded once a year. Find the interest. Show work.

Check Point 2

Use the video to find the amount of money after ten years and how much interest you would earn.

Check Point 3

A sum of $10,000 is invested at a rate of 8%. Find the balance after 5 years if the account is subject to

a. Quarterly compounding

b. Continuous compounding

Objective 2: Calculate present value.

Check Point 4

How much money should be invested today to accumulate to $10,000 in 8 years at 7%?
Show work.

Objective 3: Understand and compute effective annual yield.

Check Point 5

Find the future value and effective annual yield on $6000 in an account that pays 10%
compounded monthly. Show work.

Check Point 6

What is the effective annual yield of an account paying 8% compounded quarterly? Use the
video to answer, and show work.

SECTION 8.5

Objective 1: Determine the value of an annuity.

Check Point 1

Find the value of the annuity and the interest after three years. Include your work as shown in the video.

Check Point 2

How much will you have from the IRA after 40 years? Find the interest as shown, and round to the nearest dollar.

Check Point 3

How much would you have from the IRA when you retired? Find the interest and show work.

Objective 2: Determine regular annuity payments needed to achieve a financial goal.

Check Point 4

How much should the parents deposit each month. How much of the fund will come from deposits and how much from interest? Show work.

Objective 3: Understand stocks and bonds as investments.

Use this space to jot notes about stocks, bonds, and mutual funds.

Objective 4: Read stock tables.

52-Week High	52-Week Low	Stock	SYM	Div	Yld %	PE	Vol 100s	Hi	Lo	Close	Net Chg
63.38	42.37	Coca-Cola	CocaCl	.72	1.5	37	72032	49.94	48.33	49.50	+0.03

Objective 5: Understand accounts designed for retirement savings.

Check Point 6

a. What is the value of the 401(k) rounded to the nearest dollar after 15 years?

b. How much money would you have at 65?

c. What is the difference between the amount of money you will have accumulated in the two options?

SECTION 8.6

Objective 1: Compute the monthly payment and interest costs for a car loan.

Check Point 1

a. Find the monthly payments and total interest for Loan A.

b. Find the monthly payments and total interest for Loan B.

c. Compare the two loans.

Objective 2: Understand the types of leasing contracts.

There are no check point videos here. What is the difference between open-end and closed-end leases?

Objective 3: Understand the pros and cons of leasing versus buying a car.

Check Point 1

There are no check point videos here. What are the advantages and disadvantages of leasing?

Objective 4: Understand the different kinds of car insurance.

There are no check point videos here. What are some key things to know about auto insurance?

Objective 5: Compare monthly payments on new and used cars.

Check Point 2
What is the difference in payments for the used and new cars?

Objective 6: Solve problems related to owning and operating a car.

Check Point 3
How much would you save driving the hybrid?
If you invested that money as shown, how much would you have at the end of 7 years?

SECTION 8.7

Objective 1: Compute the monthly payment and interest costs for a mortgage.

Check Point 1
Use the video to help you answer the questions and show work; the $175,500 mortgage was financed with a 30 year fixed rate at 7.5%. Total interest over 30 years was $266,220.

a. Find the monthly payment if the time of the mortgage is reduced to 15 years.

b. Find the total interest paid over 15 years.

c. How much was saved by reducing the mortgage from 30 to 15 years?

Objective 2: Prepare a partial loan amortization schedule.

Check Point 2
Prepare the table as shown in the video.

Annual % Rate: 7.0% Amount of Mortgage: $200,000 Number of Monthly Payments: 240		Monthly Payment: $1550.00 Term: Years 20, Months 0	
Payment Number	**Interest Payment**	**Principal Payment**	**Balance of Loan**
1			
2			

Objective 3: Solve problems involving what you can afford to spend for a mortgage.

Check Point 3
Suppose your annual income is $240,000.

a. What is the maximum amount you should spend each month on a mortgage payment?

b. What is the maximum you should spend each month for total credit obligations?

c. What is the maximum you should spend monthly for all other debt?

Objective 4: Understand the pros and cons of renting versus buying.

There is no check point video here. Use this area to take notes about pros and cons of renting versus buying.

Chapter 9 Measurement

SECTION 9.1

Objective 1: Use dimensional analysis to change units of measurement.

Check Point 1
Use your video to help convert units.
a. 78 inches to feet

b. 17,160 feet to miles

c. 3 inches to yards.

Objective 2: Understand and use metric prefixes.

There is no check point video for this section. Take notes on the metric prefixes and examples as needed.

Objective 3: Convert units within the metric system.

Check Point 2
Use the video to help convert the following.

a. 8000 meters to kilometers

b. 53 meters to millimeters

c. 604 cm to hm

d. 6.72 dam to cm.

Objective 4: Convert units within the metric system.

Check Point 3
Use the video and record the answers to the following.

a.   Convert 8 feet to centimeters

b.   Convert 20 meters to yards

c.   Convert 30 meters to inches.

Check Point 4
Sixty kilometers an hour is how many mph? Show work.

SECTION 9.2

Objective 1: Use square units to measure area.

Check Point 1

Find the area of the figure as shown.

Check Point 2

Find the population density of California as shown. Record work.

Objective 2: Use dimensional analysis to change units for area.

Check Point 3

How large is 84,000,000 acres in terms of square miles? Show work.

Check Point 4

Use the video to answer the following.

a. The area of the property in hectares.

b. The price per hectare.

Objective 3: Use cubic units to measure volume.

Check Point 5
What is the volume of the region?

Objective 4: Use English and metric units to measure capacity.

Check Point 6
A pool has the volume of 10,000 cubic feet. How many gallons of water can it hold? Show the work.

Check Point 7
A fish pond holds 220,000 cubic centimeters. How many liters is this? Record the computations as shown in the video.

SECTION 9.3

Objective 1: *Apply metric prefixes to units of weight.*

There are no check point videos for this objective. Use this area to jot notes.

Objective 2: *Convert units of weight within the metric system.*

Check Point 1
Use the video to complete the conversions.

a. Convert 4.2 dg to mg.

b. Convert 620 cg to g.

Objective 3: *Use relationships between volume and weight within the metric system.*

Check Point 2
How much does the 0.145 m^3 of water weigh? Show work.

Objective 4: Use dimensional analysis to change units of weight to and from the metric system.

Check Point 3

a.   Convert 120 pounds to kilograms.

b.   Convert 500 grams to ounces.

Check Point 4

Find the man's weight in kg, and the amount of medication he should receive. Record work as shown in the video.

Objective 5: Understand temperature scales.

Check Point 4

Convert 50°C to °F. Show work.

Check Point 5

Convert 59°F to°C. Show work.

Chapter 10 Geometry

SECTION 10.1

Objective 1: Understand points, lines, and planes as a basis of geometry.

There are no check point videos for this objective. Define or describe the following.

- Points

- Lines

- Planes

Objective 2: Solve problems involving angle measures.

Check Point 1
How many degrees did the hand move on the clock?

Check Point 2
Find the measure of the angle using the diagram and the video.
$m\angle DBA$.

Check Point 3
Find the measures of the two angles; draw a diagram and show work as needed.

Check Point 4
Find the measures of the other three angles, given the first is 57°.

Objective 3: Solve problems involving angles formed by parallel lines and transversals.

Check Point 5
The measure of the given angle is 29°. Find the measures of the other angles as shown in the video. Sketch the appropriate diagram.

SECTION 10.2

Objective 1: Solve problems involving similar triangles.

Check Point 1
Label the triangle, and solve for the missing angle.

Check Point 2
Draw and label the new figure and solve for the missing angles as shown in the video.

Objective 2: Solve problems involving similar triangles.

Check Point 3
Explain why the triangles are similar, and find the missing length *x*.

Check Point 4
Find the height of the tower shown. Show work.

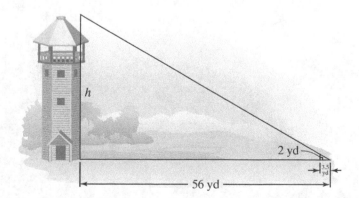

Objective 3: Solve problems using the Pythagorean Theorem.

Check Point 5
Find the length of the hypotenuse if the legs are 7 feet and 24 feet, respectively. Show work.

Check Point 6
Sketch a diagram and solve the problem using the video. A radio tower is supported by two wires that are each 130 yards long and attached to the ground 50 yards from the base of the tower. How far from the ground are the wires attached to the tower?

SECTION 10.3

| **Objective 1: Name certain polygons according to the number of sides.** |

There is no check point video for this objective. Use this space for notes about types of polygons.

| **Objective 2: Recognize the characteristics of certain quadrilaterals.** |

There are no check point videos for this objective. Record types of quadrilaterals and make a sketch of each.

| **Objective 3: Solve problems involving a polygon's perimeter.** |

Check Point 1
Find the cost to enclose the field with fencing if the field is 50 yards long and 30 yards wide, and the fencing costs $6.50 per foot. Show work.

Objective 4: *Find the sum of the measures of a polygon's angles.*

Check Point 2

Use the video to answer the following.

a. Find the sum of the measures of the angles of a 12-sided polygon.

b. Find the measure of each angle of a regular 12-sided polygon.

Objective 5: *Understand tessellations and their angle requirements.*

Check Point 3

Why is it impossible to create a tessellation using only regular octogons?

SECTION 10.4

Objective 1: Use area formulas to compute the areas of plane regions and solve applied problems.

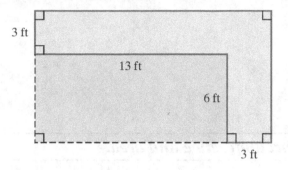

Check Point 1
Use the video to help you find the area of the path. Show work.

Check Point 2
What will it cost to carpet a rectangular floor 18 by 21 feet if the carpet costs $16 a square yard? Show work.

Check Point 3
Find the area of a parallelogram with a base of 10 inches and height of 6 inches. Sketch the appropriate diagram and note the formula used in the video.

Check Point 4
Find the area of the sail with a base of 12 feet and the height of 5 feet.

Check Point 5
Find the area of the trapezoid with bases 20 and 10 feet and the height of 7 feet. Show work.

Objective 2: Use formulas for a circle's circumference and area.

Check Point 6
Find the circumference of the circle with a diameter of 10 inches. Express the exact answer in terms of π and then approximate; round to the nearest tenth of an inch. Show work.

Check Point 7
Use the video to find how much trim is needed to go around the window.

Check Point 8
Use the video to determine the better buy: a large pizza with 18 inch diameter for $20.00 or a medium with 14 inch diameter for $14.00. Show work.

SECTION 10.5

> *Objective 1: Use volum formulas to compute the volumes of three-dimensional figures and solve applied problems.*

Check Point 1

Find the volume of the rectangular solid with length 5 feet, width 3 feet, and height 7 feet, as shown.

Check Point 2

Find the volume in cubic yards of the cube whose edges are six feet. Watch your units! Show work.

Check Point 3

A pyramid is 4 feet tall. Each side of the square base has a length of 6 feet. Find the pyramid's volume as shown in the video. Show work.

Check Point 4

Find the volume of the cylinder with a diamter of 8 inches and a height of 6 inches. Show work.

Check Point 5
Find the volume of the cone with a radius of 4 inches and a height of 6 inches. Show work.

Check Point 6
Is 350 cubic inches enough to fill the ball with a radius of 4.5 inches? Use the video to answer the question; show work.

Objective 2: Compute the surface area of a three-dimensional figure.

Check Point 7
If the length, width, and height are doubled, find the surface area of the rectangular solid as shown in the video.

SECTION 10.6

> **Objective 1: Use the lengths of the sides of a right triangle to find trigonometric ratios.**

Check Point 1

Use the video to find the sin A, cos A, and tan A with the triangle shown.

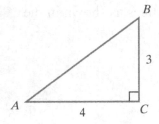

> **Objective 2: Use trigonometric ratios to find missing parts of right triangles.**

Check Points 2 & 3

Use the figures and videos to find the designated sides to the nearest centimeter. Show work.

2) a=

3) c=

> **Objective 3: Use trigonometric ratios to solve applied problems.**

Check Point 4

Use the video to help you approximate the height of the Eiffel tower to the nearest foot. Sketch the diagram.

Check Point 5

Find the angle of elevation to the nearest degree. The flagpole is 14 meters tall and casts a shadow 10 meters long. Sketch a diagram.

SECTION 10.7

Objective 1: Gain an understanding of some of the general ideas of other kinds of geometries.

Check Point 1

Use the space below to create a graph with two even and two odd vertices as shown in the video. Then describe the path that will traverse it.

Chapter 11 Counting Methods and Probability Theory

SECTION 11.1

> *Objective 1: Use the Fundamental Counting Principle to determine the number of possible outcomes in a given situation.*

Check Point 1
If there are 10 appetizers and 15 main courses, how many different meal possibilities are there?

Check Point 2
Use the video to help you rework Example 2 given that the number of non-overlapping classes of both sociology and psychology decrease by 5.

Check Point 3
How many different one-topping pizzas can be ordered?

Check Point 4
In how many ways can the car be ordered in terms of the options given?

Check Point 5

In how many ways can you answer the questions?

Check Point 6

How many different keypad sequences are there? Remember you cannot use zero as the first digit.

SECTION 11.2

> ### *Objective 1: Use the Fundamental Counting Principle to count permutations.*

Check Point 1
In how many ways can the 6 jokes be delivered if a man's is told first?

Check Point 2
In how many ways can you arrange the five books on the shelf?

> ### *Objective 2: Evaluate factorial expressions.*

Check Point 3
Evaluate without using a calculator. Show the work.
a.

b.

c.

> ### *Objective 3: Use the permutations formula.*

Check Point 4
In how many ways can they elect a president, vice-president, secretary, and treasurer?

Check Point 5

How many different programming sequence options are there if they choose 5 of the 9 comedies?

Objective 4: Find the number of permutations of duplicate items.

Check Point 6

In how many ways can the letters in the word OSMOSIS be arranged?

SECTION 11.3

Objective 1: Distinguish between permutation and combination problems.

Check Point 1
Identify which type of problem is shown.

a.

b.

Objective 2: Solve problems involving combinations.

Check Point 2
How many different combinations of pets are there, if you choose 3 of the 7?

Check Point 3
How many different 4-card hands can be dealt from 16 cards?

Check Point 4
How many five-bear collections are possible? Show work.

SECTION 11.4

Objective 1: Compute theoretical probability.

Check Point 1
Use the video to answer the questions about tossing a die.

a.

b.

c.

d.

Check Point 2
Find the probability of being dealt

a. An ace

b. A red card

c. A king

Check Point 3
Find the probability that the child would be a carrier of cystic fibrosis, but not have the disease. Show work.

Objective 2: Compute empirical probability .

Check Point 4
Find the probability using the video and figure.

a. P(the person never married)

b. P(is male)

SECTION 11.5

Objective 1: Compute probabilities with permutations.

Check Point 1

What is the probability that jokes will be told as explained in the problem? A person whose first name begins with G must be first, and a man's joke should be last.

Objective 2: Compute probabilities with combinations.

Check Point 2

Find the probability of winning the consolation prize.

Check Point 3

A club has six men and four women. If three people are selected to go to a conference, what is the chance that the group is

a. All men

b. Two men and one woman

SECTION 11.6

Objective 1: Find the probability that an event will not occur.

Check Point 1
What is the probability you are *not* dealt a diamond?

Check Point 2
If a driver is randomly selected from that data set, what is the probability that
a.   The driver is not in the 50-59 age bracket

b.   Is at least 20 years old.

Objective 2: Find the probability of one event or a second event occurring.

Check Point 3
What is the probability of rolling a 4 or 5 when tossing a die?

Check Point 4
Find the probability that a student takes math or psychology. Show work and/or a diagram as needed.

Check Point 5
Find the probability that the pointer will stop on an odd number or a number less than 5.

Check Point 6
Use the videos to solve the problem; record your work here.

a. What is the probability that the person is married or female?

b. What is the probability that the person is divorced or widowed?

Objective 3: Understand and use odds.

Check Point 7
What are the odds in favor of getting a red queen? Odds against? How is this different from probability? Explain.

Check Point 8
What are the odds against your winning the scholarship?

Check Point 9
Find the odds in favor of the horse winning. What is the probability that the horse will win?

SECTION 11.7

Objective 1: Find the probability of one event and a second event occurring.

Check Point 1
Find the probability of green on two consecutive plays on the roulette wheel.

Check Point 2
What is the probability of a family having 4 boys in a row?

Check Point 3
Use the video to solve the problems about the hurricanes in Florida.

Check Point 4
Find the probability of getting two kings from the deck of cards.

Check Point 5
Find the probability of 3 hearts.

Objective 2: Compute theoretical probabilities.

Check Point 6

Find the probability that it precedes h, given it is a vowel.

Check Point 7

Use the video to find the

a. Probability of getting a black card, given you were dealt a spade.

b. Probability of getting a spade, given you were dealt a black card.

Check Point 8

Record the probabilities as shown in the video. What implications, if any, does the information have for women?

SECTION 11.8

Objective 1: Compute expected value.

Check Point 1
Find the expected value for where the pointer will stop.

Check Point 2
Find the expected value for the number of heads using the table and the video.

Objective 2: Use expected value to solve applied problems.

Check Point 3
Rework the problem as instructed, using the video to help you. Show work.

Check Point 4
Compute the expected value of a random guess on the SAT test. Is there anything to gain or lose by guessing? What is your recommendation to a student?

Objective 3: Use expected value to determine the average payout or loss in a game of chance.

Check Point 5

What do you expect if you purchase five tickets?

Check Point 6

Find the expected value of a $1 bet and explain what it means.

Some additional explorations for probability.

1. Roll two six-sided dice 36 times and record the sums below. How does it compare to what you predicted would happen?

2. This is the Monty Hall problem. There are 3 doors. Behind two doors are goats; the third has a brand-new hybrid sports car. You choose a door. The game show host opens a different door and shows you a goat. You can now choose to stick or switch....which should you do? Why? (check your answer at Ask Dr. Math...you may be surprised! There are some great animations of this online.)

3. The ruler of a small kingdom decides he needs more boys born to increase his military force. He passes a law that dictates EVERY couple will have children until they have a boy. What do you think will happen? Simulate this by rolling a die; odds is a boy, evens is a girl. Record results for 10 families. What do you think about the results? Is it what you expected?

4. A strong wind sweeps through a department store and 30 hats are thrown on the floor. An employee rapidly picks them up and puts them on display again. However, he has no idea how they were arranged. What is the probability that all 30 would end up in their original spot?

Chapter 12 Statistics

SECTION 12.1

> ### *Objective 1: Describe the population whose properties are to be analyzed.*

Check Point 1
Does the survey seem like a good idea? Describe the population.

> ### *Objective 2: Select an appropriate sampling technique.*

Check Point 2
How would a researcher set up a random sample of the city's homeless people?

> ### *Objective 3: Organize and present data.*

Check Point 3
Use the video to help create a frequency chart for the final course grades in precalculus.

Check Point 4
Use the video to help create a grouped frequency distribution.

Check Point 5
Construct the stem/leaf plot for the data in Checkpoint 4.

Objective 4: Identify deceptions in visual displays of data.

There are no check point videos for this objective. Summarize examples of deceptions in data.

SECTION 12.2

Objective 1: Determine the mean for a data set.

Check Point 1
Find the mean for the ten highest-paid actresses.

Check Point 2
Find the mean for the frequency distribution.

Score, x	30	33	40	50
Frequency, f	3	4	4	1

Objective 2: Determine the median for a data set.

Check Point 3
Find the median for each set of data.
a.

b.

Check Point 4
Find the median for the set of data.
1, 2, 2, 2, 3, 3, 3, 3, 3, 5, 6, 7, 7, 10, 11, 13, 19, 24, 26

Check Point 5
Order the data and then find the median for the amount of type spent eating.

Check Point 6
Find the median for the frequency distribution.

x	42	43	46	51	52	54	55	56	60	61	64	69
f	1	1	1	3	1	2	2	2	1	2	1	1

Check Point 7
Find the mean and median compensation and compare. Which one is the better representation and why?

Objective 3: Determine the mode for a data set.

Check Point 8
Find the mode of the data set. 8, 6, 2, 4, 6, 8, 10, 8

Objective 4: Determine the midrange for a data set.

Check Point 9
Find the midrange for payroll for the NFL.

Check Point 10
Find the mean, median, mode, and midrange of the calories in the hot dogs.
172, 191, 182, 190, 172, 147, 146, 138, 175, 136, 179, 153, 107, 195, 135, 140, 138

Section 12.3

Objective 1: Determine the range for a data set.

Check Point 1
Find the range for 4, 2, 11, 7

Objective 2: Determine the standard deviation for a data set.

Check Points 2 & 3
Use 4, 2, 11, 7 to find first mean deviations, then standard deviation. Show work. Round to two decimal places.

Check Point 4
Find the standard deviation of both datasets. Compare.

Sample A: 73, 75, 77, 79, 81, 83
Sample B: 40, 44, 92, 94, 98, 100

Check Point 5
Which investment has a greater return? Greater risk? Which one would you pick?

SECTION 12.4

| Objective 1: *Recognize characteristics of normal distributions.* |

There are no videos for this objective; use this space to take notes about normal distributions.

| Objective 2: *Understand the 68-95-99.7 rule.* |

There are no videos for this objective; use this space to take notes.

THE 68–95–99.7 RULE FOR THE NORMAL DISTRIBUTION

1. Approximately 68% of the data items fall within 1 standard deviation of the mean (in both directions).
2. Approximately 95% of the data items fall within 2 standard deviations of the mean.
3. Approximately 99.7% of the data items fall within 3 standard deviations of the mean.

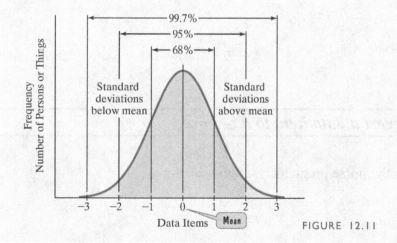

FIGURE 12.11

Objective 3:   Find scores at a specified standard deviation from the mean.

Check Point 1
The mean is 65 inches; the standard deviation is 3.5 inches. Find the height:

a.   3 standard deviations above the mean

b.   2 standard deviations below the mean.

Objective 4:   Use the 68-95-99.7 rule.

Check Point 2
Use the figure and video to find the percentage of men with heights
a.   Between 62 and 78 inches

b.   Between 70 and 78 inches

c.   Above 74 inches.

Objective 5:   Convert a data item to a z-score.

Check Point 3
Find the z-scores for the horse pregnancies. Show work.
a.

b.

c.

Check Point 4
On which test would you have a better score? Why? Explain.

Check Point 5. Find the IQ score corresponding to a z-score of

a.   −2.25

b.   1.75

Objective 6: Understand percentiles and quartiles.

Check Point 6

What does it mean to score at the 75th percentile on the SAT test?

Objective 7: Use and interpret margins of error.

Check Point 7

Find the margin of error as shown. Write a statement about the percentage of U.S. adults who read more than ten books a year. Why might some people not respond honestly to this poll?

Objective 8: Recognize distributions that are not normal.

There are no videos for this objective. Use this area to describe various distributions in your own words and diagrams.

SECTION 12.5

> ## *Objective 1: Solve applied problems involving normal distributions.*

Check Point 1

The distribution of monthly charges for cellphone plans in the United States is approximately normal with a mean of $62 and a standard deviation of $18. What percentage of plans have charges that are less than $83.60? Show work.

Check Point 2

What percentage of women have heights that exceed 69.9 inches?

Check Point 3

The refrigerators typically last an average of 14 years with a standard deviation of 2.5 years. What percentage last between 11 and 18 years?

SECTION 12.6

Objective 1: Make a scatter plot for a table of data items.

There are no videos for this objective. Use this area to describe correlation and the reading of scatterplots.

Objective 2: Interpret information given in a scatter plot.

Check Point 1
Is there a relationship between the body-mass index of daughters and the body-mass index of their mothers? Explain.

Objective 3: Compute the correlation coefficient.

Check Point 2
Determine the correlation coefficient between the number of firearms and number of deaths. What does it tell us about the strength and direction of the relationship between firearms per 100 people and deaths per 100,000 people?

Objective 4: Write the equation of the regression line.

Check Point 3
Find the equation of the regression line. How can we use it? What does it mean?

Objective 4: Use a sample's correlation coefficient to determine whether there is a correlation in the population.

Check Point 4
If you did get $r = 0.89$ for $n = 10$, then can you conclude that there is a positive correlation for all industrialized countries between firearms and deaths? Explain.

Chapter 13 Voting and Apportionment

SECTION 13.1

Objective 1: Understand and use preference tables.

Number of Votes	2100	1305	765	40
First Choice	S	A	S	B
Second Choice	A	S	A	S
Third Choice	B	B	C	A
Fourth Choice	C	C	B	C

TABLE 13.3 Preference Table for the Election of Student Body President

Check Point 1
Use the table and video to answer the questions.
a. How many students voted?

b. How many selected the candidates in this order B,S,A,C?

c. How many students selected Samir (S) as their first choice for president?

Objective 2: Use the plurality method to determine an election's winner.

Check Point 2
Who is declared the winner using the plurality method for the mayor of Smallville?

Objective 3: Use the Borda count method to determine the election's winner.

Check Point 3
Who is declared the winner using the Borda Count method?

Objective 4: Use the plurality-with-elimination method to determine an election's winner.

Check Point 4
Who is declared the winner of the mayoral election using the plurality-with-elimination method?

Objective 5: Use the pairwise comparison method to determine an election's winner.

Check Point 5
Who is the winner using the pairwise comparison method?

SECTION 13.2

Objective 1: Use the majority criterion to determine a voting system's fairness.

Check Point 1
Which candidate has the majority of votes? Which is declared the principal by the Borda count method?

Number of Votes	6	4	2	2
First Choice	A	B	B	A
Second Choice	B	C	D	B
Third Choice	C	D	C	D
Fourth Choice	D	A	A	C

Objective 2: Use the head-to-head criterion to determine a voting system's fairness.

Check Point 2
Which brand is favored? Which wins the listening test using the plurality method?

Number of Votes	3	2	2
First Choice	A	B	C
Second Choice	B	A	B
Third Choice	C	C	A

What do you think of this method?   Is it fair?

Objective 3: Use the monotonocity criterion to determine a voting system's fairness.

Check Point 3

Use the video to answer the following.

a. Using the plurality-with-elimination method, which candidate wins the first election?

b. Using the same method, who wins the second election?

c. Does this violate the monotonicity criterion? Explain.

Objective 4: Use the irrelevent alternatives criterion to determine a voting system's fairness.

Check Point 4

The election results are shown in the table. Using the pairwise comparison method, who wins the election? If B and C are eliminated, who then wins? Does this violate the irrelevent alternatives criterion? Explain.

for Mayor				
Number of Votes	**150**	**90**	**90**	**30**
First Choice	A	C	D	D
Second Choice	B	B	A	A
Third Choice	C	D	C	B
Fourth Choice	D	A	B	C

Objective 5: Understand Arrow's Impossibility Theorem.

There is no check point video for this objective. Use this area to summarize the theorem.

SECTION 13.3

Objective 1: Find standard divisors and standard quotas.

Check Point 1

Find the standard divisor and find the standard quota for each state.

State	A	B	C	D	E	Total
Population (in thousands)	1112	1118	1320	1515	4935	10,000
Standard Quota						

Objective 2: Understand the apportionment problem.

There are no check point videos for this objective. Use this area to summarize apportionment.

Objective 3: Use Hamilton's method.

Check Point 2

Use Hamilton's method to apportion the 200 seats for the data in Check Point 1.

Objective 4: Understand the quota rule.

There are no check point videos for this objective. What is the quota rule?

Objective 5: Use Jefferson's method.

Check Point 3

Use Jefferson's method with $d = 49.3$ to apportion the 200 congressional seats.

Objective 6: Use Adam's method.

Check Point 4

Use Adam's method; start with $d = 50.5$ and apportion the 200 seats.

Objective 7: Use Webster's method.

Check Point 4

Use Webster's method with $d = 49.8$ and apportion the 200 seats.

SECTION 13.4

Objective 1: Understand and illustrate the Alabama paradox.

Check Point 1
Use Hamilton's method to show the Alabama paradox occurs if the number of congressional seats increases from 99 to 100.

State	A	B	C	D	Total
Population	2060	2080	7730	8130	20,000

Objective 2: Understand and illustrate the population paradox.

Check Point 1
A small country has 100 seats in its congress, divided among three states according to populations. The table shows each state's population before and after the country's population increase. Use the video and table to answer the questions below.

TABLE 13.47				
State	A	B	C	Total
Original Population	19,110	39,090	141,800	200,000
New Population	19,302	39,480	141,800	200,582

a. Use Hamilton's method to apportion the 100 seats using the original population.

b. Find the percentage increase in populations A and B.

c. Use Hamilton's method to apportion the 100 seats using the new population. Show that the paradox occurs.

Objective 3: Understand and illustrate the new-states paradox.

Check Point 3
Use the video to answer the questions. Show work.

a. Use Hamilton's method to apportion the counselors to the two schools. The two high schools have enrollments of 2574 and 9426 students, and the counseling staff numbers 100.

b. Suppose a new high school with 750 students is added to the district. The district hires 6 new counselors. Use Hamilton's method to show the new-states paradox occurs when the counselors are re-apportioned.

Objective 4: Understand Balinski and Young's Impossibility Theorem.

There are no check point videos for this objective. Summarize the impossibility theorem here. What impact does that have on elections?

Chapter 14 Graph Theory

SECTION 14.1

Objective 1: *Understand relationships in a graph.*

Check Point 1

Explain why the two graphs are equivalent.

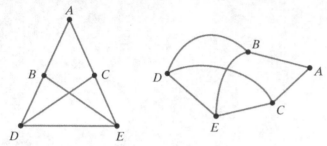

Objective 2: *Model relationships using graphs.*

Check Point 2

Draw a graph that models the layout of Metroville.

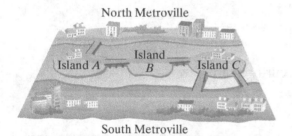

Check Point 3
Create a graph that models the relationship among the states.

Check Point 4
Draw a graph to model the floor plan.

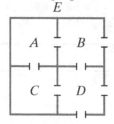

Check Point 5
Draw a graph to model the streets walked by the security guard.

Objective 3: Understand and use the vocabulary of graph theory.

Check Point 6
List the pairs of adjacent vertices for the graph.

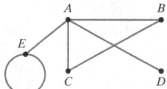

SECTION 14.2

Objective 1: *Understand the definition of an Euler path.*

Objective 2: *Understand the definition of an Euler circuit.*

There are no videos for this objective. Use this space to define:

Euler path:

Euler circuit:

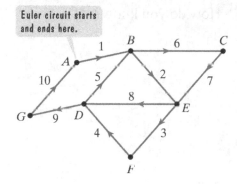

Objective 3: *Use Euler's Theorem.*

Check Point 1

Find and name an Euler path that starts at E and ends at D for the diagram.

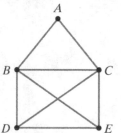

Check Point 2

Find a circuit that starts and ends at G.

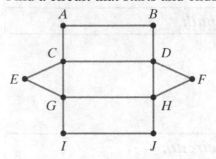

Objective 4: Solve problems using Euler's Theorem.

Check Point 3

Is it possible to find a path that uses each door exactly once? How do you know?

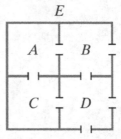 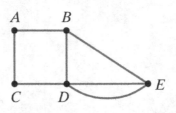

Objective 5: Use Fleury's Algorithm to find possible Euler paths and circuits.

Check Point 4

Find a circuit by using Fleury's Algorithm.

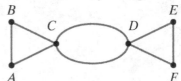

SECTION 14.3

> **Objective 1: Understand the definitions of Hamilton paths and Hamilton circuits.**

Check Point 1
Find a Hamilton path and a Hamilton circuit that begin at E.

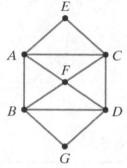

What is the key difference between Hamilton paths and circuits and Euler paths and circuits?

> **Objective 2: Find the number of Hamilton circuits in a complete graph.**

Check Point 2
Determine the number of Hamilton circuits in a complete graph with
a. Three vertices

b. Six vertices

c. Ten vertices.

> **Objective 3: Understand and use weighted graphs.**

Check Point 3
Find the cost of the circuit A, C, B, D, A.

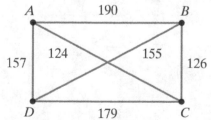

Objective 4: Use the Brute Force Method to solve traveling salesperson problems.

Check Point 4

Use the Brute Force Method to find the optimal solution.

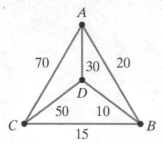

Objective 5: Use the Nearest Neighbor Method to approximate solutions to traveling salesperson problems.

Check Point 5

Begin at A, use the Nearest Neighbor Method, and find the total weight of the circuit created.

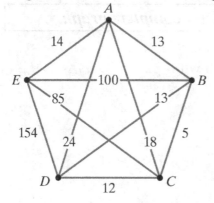

SECTION 14.4

Objective 1: *Understand the definition and properties of a tree.*

Check Point 1

Which graph is a tree? Explain why each graph is/is not a tree.

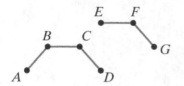 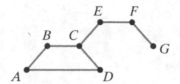 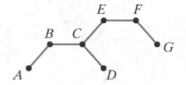

Objective 2: *Find a spanning tree for a connected graph.*

Check Point 2

Find a spanning tree for the graph shown.

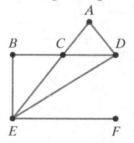

Objective 3: Find the minimum spanning tree for a weighted graph.

Check Point 3

Use Kruskal's algorithm to find the minimum spanning tree and find the total weight.

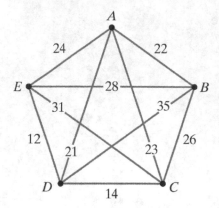

Video 1: Introduction to MathTalk

Summary

In this video, the concepts of linear equations and their graphs are introduced. We show the process of creating a line by plotting the average temperature for six months in Chicago. The line allows the presenter to derive the formula that shows the temperature increasing by an average of ten degrees each month. The process of creating these types of formulas from a set of data is called modeling. The presenter explains that formulas can be applied to real-world examples such as a doctor calculating the risk for cancer or an engineer calculating the height of a bridge from land.

Class discussion questions

1. What other types of careers can you think of that use formulas in their work? Give an example.

2. Within the video, the presenter mentions that all the formulas in textbooks have real uses. For example, both the x and y values could mean something like cancer or the protein in the blood. What other equations can you think of that most people use in everyday life?

3. Do all equations with a real life meaning have to be connected straight lines? What examples can you think of which demonstrate either a connected or disconnected line?

4. The presenter mentions that saying, "December would be 130 degrees" is called extrapolation. What do you think he means by extrapolation?

466

Video 2: Exponentiation

Summary

This video opens with a large mobile created from an assortment of toys. The presenter uses this concept of a mobile to explain exponentiation. He discusses how one object becomes two, two objects become four, four objects become eight, and so on. He explains how exponentials work by folding a newspaper and discussing how many folds are needed to reach the top of the Woolworth Tower. While walking on Wall Street, a conversation is held with a man playing a guitar about how exponentials are used in his career to determine how much commodities will be worth over time.

Class discussion questions

1. What are some of the similarities and differences of multiplication and exponentiation?

2. Which is larger, 10^x or x^{10}?

3. Why do you think many people confuse exponentials and think that it would really take thousands and millions of fold to reach the top of the Woolworth tower?

Video 3: Introduction to Trigonometry

Summary

This video focuses on how to convert angles to distances by using trigonometric functions. The presenter begins in a park measuring the sides and angles of the cardboard skyscrapers that were built by a group of people. With discussions about corresponding sides and angles, a conclusion is drawn that larger angles have larger corresponding sides. The presenter explains the Pythagorean Theorem – that the sum of the legs squared is equal to the hypotenuse squared. One example used to explain this concept is the height of a window on a house.

Class discussion questions

1. What kind of triangle does the Pythagorean Theorem apply to?

2. You all should remember special triangles such as 30-60-90 and 45-45-90 triangles. Use the trigonometric functions to show that the side lengths are correct.

3. What are some other instances in which a person may need to use Trigonometric functions?

4. The mnemonic of SOHCAHTOA is expanded to "Stephen, Oh Heck, Crocodiles And Hedgehogs Took Our Apples." What other words can you think of to help remember this?

468

Video 4: Logarithms

Summary

In this video, the presenter explains that logarithms are used to turn multiplication into addition and exponentiation into multiplication. A conversation is held with a traffic engineer about how traffic times involve logs and follows a logarithmic curve describing how often a journey will take a certain amount of time. The traffic engineer explains how having a delay due to rubbernecking as well as heavy traffic will multiply together to create a longer wait time.

Class discussion questions

1. Are all logs in base 10? What other log base can you think of that it used most often?

2. Describe a relationship between an equation in logarithmic form and an equivalent equation in exponential form.

3. Redraw the mobile that was drawn on the sidewalk in the video. Using this, determine how you find 32 x 8.

Video 5: Pascal's Triangle (Part 1)

Summary

This video uses a bean machine, also known as a quincunx, to demonstrate both the central limit theorem as well as Pascal's Triangle. Students drop marbles from the top of the quincunx and see which slot they land in at the bottom. After numerous marbles have been dropped, it is observed that a bell curve forms and the presenter explains that this happens any time random processes occur. This reasoning helps explain how Pascal's Triangle is formed, where each number represents how many paths the marble can take around that peg. All the numbers within the triangle also have different meanings, such as the rows representing the powers of 11, adding the rows giving the powers of 2, and the diagonals representing how shapes are organized in space.

Class discussion questions

1. Using what you know about Pascal's Triangle, what is $(a+b)^4$?

2. The quincunx is said to represent how random processes occur and that it will result in the bell curve. What does the presenter mean by a bell curve and when does this become more accurate?

3. At the end of the video, the presenter mentions that the diagonals represent how shapes are organized in space. What do you think he means by this?

470

Video 6: Pascal's Triangle (Part 2)

Summary

Going along with the previous video on Pascal's Triangle, the presenter goes into more detail on the diagonals of the triangle. He first explains that in order to create the triangle, you have to start with a triangle of ones and then add up two numbers on one row to create a new number on the row after. How shapes are organized in space is the basis for the diagonals of this triangle. It is explained that the third diagonal (1, 3, 6, 10, etc.) are the triangle numbers. The fourth diagonal numbers (1, 4, 10, 20, etc.) are the tetrahedral numbers of the third dimension. The fifth diagonal, however, can be described using the intersection points of different shapes. Fractals are then described by shading in the even numbers within Pascal's Triangle to form the Sierpinski Triangle, a series of triangles within each other.

Class discussion questions

1. What is Pascal's triangle? How do you find the numbers in any row of the triangle? Describe some patterns you notice within the triangle.

2. It is discussed that the entries in each diagonal represents different properties of shapes. What do you notice when you add up the numbers in each diagonal?

3. The Sierpinski Triangle is created by a series of triangles within each other. These shapes are called fractals. What other fractals can you think of?

Video 7: Combinations, Permutations, and Pascal's Triangle

Summary

This video utilizes a soccer team and a band to help understand how combinations and permutations work. Coaches sometimes assign very specific roles, such as left midfield, right midfield, and center midfield, while others just say you three go play midfield. The presenter is able to explain how this mathematically creates a large difference. Allowing the players to choose their specific positions on an 11 player team changes the possibilities from 11 x 10 x 9 x 8 to (11 x 10 x 9 x 8)/(4 x 3 x 2 x 1). The presenter then explains that Pascal's Triangle can also be used to calculate combinations. For example, to find out what choosing 2 from 4 is, count down 4 rows then over 2 numbers. The same simple processes of adding numbers together produce all sorts of things that we can see around us in the world.

Class discussion questions

1. The presenter gives the formula for combinations as $_nC_r = (n!)/((n-r)!r!)$. Using this formula, what is 5 choose 2?

2. Explain what the difference is between a combination and a permutation is.

3. It's often said that one of the most difficult hands to get in poker is a royal flush consisting of a 10, Jack, Queen, King, and Ace of the same suit. Using the information about combinations, what is the probability that you will get dealt a royal flush?

472

Video 8: Introduction to *e*

Summary

In this video, the presenter explains interest in banking by loaning milk to another individual. She asks to borrow a gallon of milk for an hour at a 100% interest rate. Then she wants to borrow those two gallons for just a half hour. This process of reinvestment continues a few more times, which is enough for the presenter to be able to explain that if the milk was reinvested an infinite number of times, the price would eventually reach *e*. *e* is the universal number that signifies both growth and decay. The presenter also notes that he uses logarithms to base *e* in his line of work as a medical statistician.

Class discussion questions

1. If you missed a credit card payment of $254.20 and the interest rate on the card is 14.95%, how much will be due next month?

2. The concept of the milk being reinvested an infinite number of times is called compounding continuously, the formula for which is $A=Pe^{rt}$, given that A is the balance after t years in an account with principal P. If your parents invest $5,000 the day you are born into a college fund with an interest rate of 9.5%, how much money will they have for you to go to college at 18?

3. What situations can you think of that compounding monthly will benefit you more than compounding continuously?

Video 9: Linear and Non-Linear Relationships

Summary

In this video, we use multiple real world examples to explain differences between linear and non-linear relationships. Two examples are given using various people found walking around them. In the first example, a curve was found involving people jumping as high as they can and marking within their age group. A linear relationship was found when the height of a person was traced along with their corresponding shoe size. The presenter interviewed multiple people after running the Boston marathon and found that female times have a linear relationship but male times often have a curvature relationship. The presenter discusses the differences between linear, quadratic, and cubic functions. He states that generally when there are more terms in a function, there are more changes to the graph.

Class discussion questions

1. Does real life operate more on a linear relationship or non-linear relationship?

2. What does it mean to be a linear or non-linear relationship? What is a relation?

3. The study of the time it takes to run a marathon was interesting because females and males differed as they got older. Why do you think that men and women ran at a different relation?

4. Can these relationships be used to explain every situation? Do they explain a whole population?

474

Video 10: Conic Sections and Ellipses

Summary

This video opens with a discussion on how biceps are essentially two cones on top of each other. To measure the mass, you simply calculate the volume of the two cones. A theory is that all circles are made up of short straight lines so that when you place a marble in the center of a circle and flick towards the edge, it will always return to the center. With an ellipse, there are two centers; if you flick a marble from one of those centers to the edge it will return to the opposite center. These centers within the ellipse are called foci. This concept is used with a lithotripter in medicine to break kidney stones. Place the sound generator at one foci and the kidney stone at the other. If the sound goes off, it will transfer to the other foci to break the kidney stone.

Class discussion questions

1. What is an ellipse and how does it mainly differ from a circle?

2. The video explained that kidney stones are broken through the use of an elliptic lithotripter. You and a friend are at a museum that happens to be in the shape of an ellipse. If your friend was at one focus of the room and you were at the other, would your conversation be clearer than if you were somewhere else in the room?

3. What other things do you notice that are in the shape of an ellipse instead of a circle? Why do you think they are shaped this way?

Video 11: Applications of *e*

Summary

In this video, the presenter reminds the audience about the importance of *e*, the universal constant of growth. He presents multiple variations of the exponential graph, including that of decay and human population growth. The presenter is back at the finish line of the Boston marathon asking runners for their age and the time it took them to finish. After graphing the data he received, it results in the normal distribution which leads to a discussion on how *e* is used with the normal curve. A discussion with a doctor demonstrates that Euler's constant is also used within the medical field to help remove tumors as well as in x-rays and ultrasounds.

Class discussion questions

1. What other examples can you think of that model exponential growth or decay?

2. Why do you think that the age of the marathon runners models a bell or normal curve?

3. One problem with all exponential growth models is that nothing can grow exponentially forever. Describe factors that might limit the size of a population.

476

Video 12: Phi and the Golden Ratio

Summary

In this video, the presenter discusses the constant phi, 1.618, which is often called the golden number. Phi is often used to describe the dimensions of human beauty, as it was used in creating the Parthenon and painting Mona Lisa. The golden ratio is explained through the drawing of a rectangle off a square and has a special property that the ratio of phi to one is the same as phi plus one to phi. Phi also relates to the Fibonacci sequence such that if you divide two subsequent numbers you will get closer to phi as the numbers get larger. The Fibonacci sequence is seen in the real world through petals of a rose and seeds in a sunflower.

Class discussion questions

1. Using the proportion set up by the golden rectangle and the quadratic formula, show that phi is in fact a solution to the quadratic equation.

2. The presenter mentions that the Golden Ratio occurs frequently in nature and the way things are created. Think of a few things you use daily that are created using phi.

3. What is the connection between the Fibonacci sequence and the golden ratio?

Video 13: Sequences

Summary

In this video, sequences are presented in a variety of ways using the topics of multiple previous videos. The presenter brings up compound interest with milk, the Basel problem and harmonic series, phi and the Fibonacci sequence, and bacterial growth. He then uses the graph of $y = x^2$ to try and locate the slope when $x = 5$. The way in which one calculates the slope at a particular point is exactly how the constants e, pi, and phi were created, by imagining a sequence that goes on forever. The relation of the slope of a point is that a curve is essentially really tiny straight lines put together. The presenter concludes the video with finding the area under the curve as a series of smaller and smaller trapezoids.

Class discussion questions

1. The presenter mentions to imagine multiple sequences that go on forever. Do all sequences continue or will some come to an end?

2. Think about some different sequences mentioned within the video or some of your own. Are all of the sequences obtained through the same mathematical operations?

3. The Fibonacci sequence is brought up within the video. What type of sequence is this?

CHAPTER 1 – Problem Solving and Critical Thinking

Guided Practice:

☐ Review each of the following **Solved Problems** and complete each **Pencil Problem**.

Objective 1.R.1: Translate English phrases into algebraic expressions	
✔ **Solved Problem #1**	✎ **Pencil Problem #1** ✎
1. Write each English phrase as an algebraic expression. Let the variable x represent the number.	1. Write each English phrase as an algebraic expression. Let the variable x represent the number.
1a. the product of 6 and a number $6x$	**1a.** four more than a number
1b. a number added to 4 $4 + x$	**1b.** nine subtracted from a number
1c. three times a number, increased by 5 $3x + 5$	**1c.** three times a number, decreased by 5

480

1d. twice a number subtracted from 12

$12 - 2x$

1d. one less than the product of 12 and a number

1e. the quotient of 15 and a number

$\dfrac{15}{x}$

1e. six more than the quotient of a number and 30

Objective 1.R.2: Find the coordinates of points in the rectangular coordinate system

✔ *Solved Problem #2*

2. Determine the coordinates of points *E*, *F*, and *G*.

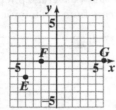

From the origin, point *E* is left 4 units and down 2 units.
Coordinates: $E(-4,-2)$

From the origin, point *F* is left 2 units.
Coordinates: $F(-2,0)$

From the origin, point *G* is right 6 units.
Coordinates: $G(6,0)$

✎ *Pencil Problem #2*✎

2. Determine the coordinates of points *A*, *C*, and *E*.

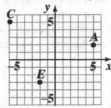

Objective 1.R.3: Evaluate algebraic expressions

✔ *Solved Problem #3*

3. Divorce rates are considerably higher for couples who marry in their teens. The line graphs in the figure show the percentages of marriages ending in divorce based on the wife's age at marriage.

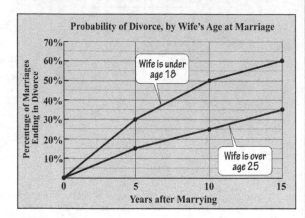

The following mathematical models approximate the data displayed by the line graphs.

Wife is under 18 at time of marriage:
$$d = 4n + 5$$

Wife is over 25 at time of marriage:
$$d = 2.3n + 1.5$$

3a. Use the appropriate formula to determine the percentage of marriages ending in divorce after 15 years when the wife is under 18 at the time of marriage.

$$d = 4n + 5$$
$$d = 4(15) + 5 = 65$$

65% of marriages end in divorce after 15 years when the wife is under 18 at the time of marriage

✎ *Pencil Problem #3* ✎

3. The bar graph shows average price of a movie ticket for selected years from 1980 through 2010.

A mathematical model that approximates the data displayed by the bar graph is

$$T = 0.15n + 2.72$$

where T represents the average movie ticket price and n is number of years after 1980.

3a. Use the formula to find the average ticket price 10 years after 1980, or in 1990. Does the mathematical model underestimate or overestimate the average ticket price shown by the bar graph for 1990? By how much?

3b. Use the appropriate line graph in the figure to determine the percentage of marriages ending in divorce after 15 years when the wife is under 18 at the time of marriage.

According to the line graph, 60% of marriages end in divorce after 15 years when the wife is under 18 at the time of marriage.

3b. Does the mathematical model underestimate or overestimate the average ticket price shown by the bar graph for 2010? By how much?

3c. Does the value given by the mathematical model underestimate or overestimate the actual percentage of marriages ending in divorce after 15 years as shown by the graph? By how much?

$65\% - 60\% = 5\%$

The mathematical model overestimates the actual percentage shown in the graph by 5%.

484

Objective 1.R.4: Use the order of operations agreement

✔ *Solved Problem #4*	✎ *Pencil Problem #4*
4a. Simplify: $20 + 4 \cdot 3 - 17$	**4a.** Simplify: $7 + 6 \cdot 3$

$$20 + 4 \cdot 3 - 17 = 20 + 12 - 17$$
$$= 20 + 12 - 17$$
$$= 15$$

4b. Simplify: $3 \cdot 2^2$

$$3 \cdot 2^2 = 3 \cdot 4$$
$$= 12$$

4b. Simplify: $(4 \cdot 5)^2 - 4 \cdot 5^2$

4c. Simplify: $7^2 - 48 \div 4^2 \cdot 5 - 2$

$$7^2 - 48 \div 4^2 \cdot 5 - 2 = 49 - 48 \div 16 \cdot 5 - 2$$
$$= 49 - 3 \cdot 5 - 2$$
$$= 49 - 15 - 2$$
$$= 34 - 2$$
$$= 32$$

4c. Simplify: $8^2 - 16 \div 2^2 \cdot 4 - 3$

4d. Simplify: $4[3(6 - 11) + 5]$

$$4[3(6 - 11) + 5] = 4[3(-5) + 5]$$
$$= 4[-15 + 5]$$
$$= 4[-10]$$
$$= -40$$

4d. Simplify: $2[5 + 2(9 - 4)]$

4e. Simplify: $25 \div 5 + 3[4 + 2(7-9)^3]$

$$
\begin{aligned}
25 \div 5 + 3[4 + 2(7-9)^3] &= 25 \div 5 + 3[4 + 2(-2)^3] \\
&= 25 \div 5 + 3[4 + 2(-8)] \\
&= 25 \div 5 + 3[4 - 16] \\
&= 25 \div 5 + 3[-12] \\
&= 5 + (-36) \\
&= -31
\end{aligned}
$$

4e. Simplify: $[7 + 3(2^3 - 1)] \div 21$

4f. Simplify: $\dfrac{5(4-9) + 10 \cdot 3}{2^3 - 1}$

$$
\begin{aligned}
\frac{5(4-9) + 10 \cdot 3}{2^3 - 1} &= \frac{5(-5) + 10 \cdot 3}{8 - 1} \\
&= \frac{-25 + 30}{7} \\
&= \frac{5}{7}
\end{aligned}
$$

4f. Simplify: $\dfrac{(-11)(-4) + 2(-7)}{7 - (-3)}$

4g. Evaluate $-x^2 - 4x$ for -5.

$$
\begin{aligned}
-x^2 - 4x &= -(-5)^2 - 4(-5) \\
&= -25 + 20 \\
&= -5
\end{aligned}
$$

4g. Evaluate $x^2 + 5x$ for $x = 3$.

486

1a. $x + 4$ **1b.** $x - 9$ **1c.** $3x - 5$ **1d.** $12x - 1$ **1e.** $\dfrac{x}{30} + 6$ **2.** $A\ (5,2),\ C\ (-6,5),\ E(-2,-3)$

3a. \$4.22; underestimates by \$0.01   **3b.** underestimates by \$0.63

4a. 25   **4b.** 300   **4c.** 45   **4d.** 30   **4e.** $\dfrac{4}{3}$ **4f.** 3   **4g.** 24

CHAPTER 2 – Set Theory

Guided Practice:

☐ Review each of the following *Solved Problems* and complete each *Pencil Problem*.

Objective 2.R.1: Understand and use inequality symbols	
✔ *Solved Problem #1*	✎ *Pencil Problem #1* ✎
1a. Insert either $<$ or $>$ to make the statement true. $\qquad -19 \quad -6$ Since -19 is to the left of -6 on the number line, then $-19 < -6$.	**1a.** Insert either $<$ or $>$ to make the statement true. $\qquad -\pi \quad -3.5$
1b. Determine if the inequality is true or false. $\qquad -2 \geq -2$ Because $-2 = -2$ is true, then $-2 \geq -2$ is true.	**1b.** Determine if the inequality is true or false. $\qquad 0 \geq -6$
1c. Determine if the inequality is true or false. $\qquad -4 \geq 1$ Because neither $-4 > 1$ nor $-4 = 1$ is true, then $-4 \geq 1$ is false.	**1c.** Determine if the inequality is true or false. $\qquad -17 \geq 6$

488

Answers for Pencil Problems:

1a. $-\pi > -3.5$ **1b.** true   **1c.** false

CHAPTER 5 – Number Theory and the Real Number System

Guided Practice:

☐ Review each of the following *Solved Problems* and complete each *Pencil Problem*.

Objective 5.R.1: Evaluate exponential expressions

✔ *Solved Problem #1*	✎ *Pencil Problem #1* ✎
1a. Evaluate: 6^2	**1a.** Evaluate: 9^2
$6^2 = 6 \cdot 6$ $\quad = 36$	
1b. Evaluate: $(-1)^4$	**1b.** Evaluate: $(-4)^3$
$(-1)^4 = (-1)(-1)(-1)(-1)$ $\qquad = 1$	
1c. Evaluate: -1^4	**1c.** Evaluate: $(-5)^4$
$-1^4 = -(1 \cdot 1 \cdot 1 \cdot 1)$ $\qquad = -1$	

490

✔ *Solved Problem*	✏ *Pencil Problem #2* ✏
2a. Solve and check: $\dfrac{x}{3} = 12$	**2a.** Solve and check: $\dfrac{x}{6} = 5$

$$\frac{x}{3} = 12$$
$$3 \cdot \frac{x}{3} = 12 \cdot 3$$
$$1x = 36$$
$$x = 36$$

The solution set is $\{36\}$.

Check: $\dfrac{x}{3} = 12$

$$\frac{36}{3} = 12$$
$$12 = 12, \text{ true}$$

2b. Solve: $-11y = 44$ | **2b.** Solve: $-28 = 8z$

$$-11y = 44$$
$$\frac{-11y}{-11} = \frac{44}{-11}$$
$$1x = -4$$
$$x = -4$$
The solution set is $\{-4\}$.

2c. Solve: $\dfrac{2}{3}y = 16$ | **2c.** Solve: $28 = -\dfrac{7}{2}x$

$$\frac{2}{3}y = 16$$
$$\frac{3}{2}\left(\frac{2}{3}y\right) = \frac{3}{2} \cdot 16$$
$$1y = 24$$
$$y = 24$$
The solution set is $\{24\}$.

491

Objective 5.R.3: Find square roots

✔ *Solved Problem*	✐ *Pencil Problem #3* ✐

3a. Evaluate: $\sqrt{81}$

$\sqrt{81} = 9$
The principal square root of 81 is 9.

3a. Evaluate: $\sqrt{36}$

3b. Evaluate: $\sqrt{\dfrac{1}{25}}$

$\sqrt{\dfrac{1}{25}} = \dfrac{1}{5}$ because $\left(\dfrac{1}{5}\right)^2 = \dfrac{1}{25}$.

3b. Evaluate: $\sqrt{\dfrac{1}{9}}$

3c. Evaluate: $\sqrt{36+64}$

$\sqrt{36+64} = \sqrt{100} = 10$

3c. Evaluate: $\sqrt{33-8}$

3d. Evaluate: $\sqrt{36} + \sqrt{64}$

$\begin{aligned}\sqrt{36} + \sqrt{64} &= 6+8\\ &= 14\end{aligned}$

3d. Evaluate: $\sqrt{144} + \sqrt{25}$

492

✔ *Solved Problem #3*

🖉 *Pencil Problem #4* 🖉

4. The mathematical model $P = 2.2\sqrt{t} + 45$ describes the percentage of bachelor's degrees, P, awarded to women in U.S. colleges t years after 1975.
    Use the formula to find the percentage, to the nearest percent, of degrees awarded to women in 1995.

1995 is 20 years after 1975. Therefore, substitute 20 for t into the formula.

$$P = 2.2\sqrt{t} + 45$$
$$P = 2.2\sqrt{20} + 45 \approx 55$$

According to the model, about 55% of degrees were awarded to women in 1995.

4. Use a calculator to approximate the expression. Round to three decimal places.
    $$\frac{-5 + \sqrt{321}}{6}$$

Answers for Pencil Problems

1a. 81　**1b.** −64　**1c.** 625　**2a.** {30}　**2b.** $\left\{-\dfrac{7}{2}\right\}$　**2c.** {−8}　**3a.** 6　**3b.** $\dfrac{1}{3}$　**3c.** 5　**3d.** 17　**4.** 2.153

CHAPTER 6 – Algebra:
Equations and Inequalities

Guided Practice:

☐ Review each of the following **Solved Problems** and complete each **Pencil Problem**.

Objective 6.R.1: Determine whether a number is a solution of an equation

✔ **Solved Problem #1**

1a. Determine whether the given number is a solution of the equation.
$9x - 3 = 42; \ 6$

$9x - 3 = 42$
$9(6) - 3 = 42$
$54 - 3 = 42$
$\quad 51 = 42, \ \text{false}$

6 is not a solution.

✎ **Pencil Problem #1**✎

1a. Determine whether the given number is a solution of the equation.
$5a - 4 = 2a + 5; \ 3$

1b. Determine whether the given number is a solution of the equation.
$2(y + 3) = 5y - 3; \ 3$

$2(y + 3) = 5y - 3$
$2(3 + 3) = 5(3) - 3$
$\quad 2(6) = 15 - 3$
$\quad\quad 12 = 12, \ \text{true}$

3 is a solution.

1b. Determine whether the given number is a solution of the equation.
$2(w + 1) = 3(w - 1); \ 7$

Objective 6.R.2: Use the addition property of equality to solve equations	
✔ **Solved Problem #2**	✎ **Pencil Problem #2** ✎
2a. Solve and check: $x - 5 = 12$	**2a.** Solve and check: $x - 4 = 19$

2a. Solve and check: $x - 5 = 12$

$$x - 5 = 12$$
$$x - 5 + 5 = 12 + 5$$
$$x + 0 = 17$$
$$x = 17$$

The solution set is $\{17\}$.

Check: $x - 5 = 12$
$$17 - 5 = 12$$
$$12 = 12, \text{ true}$$

2a. Solve and check: $x - 4 = 19$

2b. Solve and check: $z + 2.8 = 5.09$

$$z + 2.8 = 5.09$$
$$z + 2.8 - 2.8 = 5.09 - 2.8$$
$$z + 0 = 2.29$$
$$z = 2.29$$

The solution set is $\{2.29\}$.

Check: $z + 2.8 = 5.09$
$$2.29 + 2.8 = 5.09$$
$$5.09 = 5.09, \text{ true}$$

2b. Solve and check: $7 + z = 11$

496

2c. Solve: $-\dfrac{1}{2} = x - \dfrac{3}{4}$

$$-\frac{1}{2} = x - \frac{3}{4}$$

$$-\frac{1}{2} + \frac{3}{4} = x - \frac{3}{4} + \frac{3}{4}$$

$$-\frac{2}{4} + \frac{3}{4} = x$$

$$\frac{1}{4} = x$$

The solution set is $\left\{\dfrac{1}{4}\right\}$.

2c. Solve: $t + \dfrac{5}{6} = -\dfrac{7}{12}$

2d. Solve and check: $8y + 7 - 7y - 10 = 6 + 4$

$$8y + 7 - 7y - 10 = 6 + 4$$

$$y - 3 = 10$$

$$y - 3 + 3 = 10 + 3$$

$$y = 13$$

The solution set is $\{13\}$.

Check:
$$8y + 7 - 7y - 10 = 6 + 4$$
$$8(13) + 7 - 7(13) - 10 = 6 + 4$$
$$104 + 7 - 91 - 10 = 10$$
$$111 - 101 = 10$$
$$10 = 10, \text{ true}$$

2d. Solve and check: $6y + 3 - 5y = 14$

2e. Solve and check: $7x = 12 + 6x$

$$7x = 12 + 6x$$
$$7x - 6x = 12 + 6x - 6x$$
$$x = 12$$

The solution set is $\{12\}$.

Check:
$$7x = 12 + 6x$$
$$7(12) = 12 + 6(12)$$
$$84 = 12 + 72$$
$$84 = 84, \text{ true}$$

2e. Solve and check: $12 - 6x = 18 - 7x$

2f. Solve and check: $3x - 6 = 2x + 5$

$$3x - 6 = 2x + 5$$
$$3x - 2x - 6 = 2x - 2x + 5$$
$$x - 6 = 5$$
$$x - 6 + 6 = 5 + 6$$
$$x = 11$$

The solution set is $\{11\}$.

Check:
$$3x - 6 = 2x + 5$$
$$3(11) - 6 = 2(11) + 5$$
$$33 - 6 = 22 + 5$$
$$27 = 27, \text{ true}$$

2f. Solve and check: $4x + 2 = 3(x - 6) + 8$

498

✔ *Solved Problem #3*

✎ *Pencil Problem #3*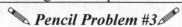

3a. Write the sentence as an equation. Let the variable x represent the number.

The quotient of a number and 6 is 5.

$$\frac{x}{6} = 5$$

3a. Write the sentence as an equation. Let the variable x represent the number.

Four times a number is 28.

3b. Write the sentence as an equation. Let the variable x represent the number.

Seven decreased by twice a number yields 1.

$$7 - 2x = 1$$

3b. Write the sentence as an equation. Let the variable x represent the number.

Five times a number is equal to 24 decreased by the number.

Objective 6.R.4: Solve linear equations	
✔ *Solved Problem #4*	✎ *Pencil Problem #4*✎

4a. Solve: $-7x + 25 + 3x = 16 - 2x - 3$

Simplify the algebraic expression on each side.
$-7x + 25 + 3x = 16 - 2x - 3$
$$-4x + 25 = 13 - 2x$$

Collect variable terms on one side and constant terms on the other side.
$-4x + 25 = 13 - 2x$
$-4x + 25 + 2x = 13 - 2x + 2x$
$$-2x + 25 = 13$$
$-2x + 25 - 25 = 13 - 25$
$$-2x = -12$$

Isolate the variable and solve.
$$\frac{-2x}{-2} = \frac{-12}{-2}$$
$$x = 6$$

The solution set is $\{6\}$.

4a. Solve: $4x - 9x + 22 = 3x + 30$

4b. Solve: $4(2x + 1) - 29 = 3(2x - 5)$

Simplify the algebraic expression on each side.
$4(2x + 1) - 29 = 3(2x - 5)$
$$8x + 4 - 29 = 6x - 15$$
$$8x - 25 = 6x - 15$$

Collect variable terms on one side and constant terms on the other side.
$8x - 6x - 25 = 6x - 6x - 15$
$$2x - 25 = -15$$
$2x - 25 + 25 = -15 + 25$
$$2x = 10$$

Isolate the variable and solve.
$$\frac{2x}{2} = \frac{10}{2}$$
$$x = 5$$

The solution set is $\{5\}$.

4b. Solve: $5(2x - 8) - 2 = 5(x - 3) + 3$

500

✔ *Solved Problem #5*   ✎ *Pencil Problem #5✎*

5a. Graph: −4

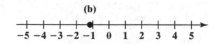

5a. Graph: 2

5b. Graph: −1.2

5b. Graph: $-\dfrac{16}{5}$

Objective 6.R.6: Combine like terms	
✔ *Solved Problem #6*	✎ *Pencil Problem #6*✎

6a. Combine like terms: $7x + 3x$

$7x + 3x = (7 + 3)x$
$\qquad\quad = 10x$

6a. Combine like terms: $7x + 10x$

6b. Combine like terms: $9a - 4a$

$9a - 4a = (9 - 4)a$
$\qquad\quad = 5a$

6b. Combine like terms: $11a - 3a$

6c. Simplify: $9x + 6y + 5x + 2y$

$9x + 6y + 5x + 2y = 9x + 5x + 6y + 2y$
$\qquad\qquad\qquad\quad = (9 + 5)x + (6 + 2)y$
$\qquad\qquad\qquad\quad = 14x + 8y$

6c. Simplify:

502

Objective 6.R.7: Multiply Monomials

✔ Solved Problem #7

7. Multiply: $(-5x^4)(4x^5)$

$$(-5x^4)(4x^5) = (-5 \cdot 4)(x^4 \cdot x^5)$$
$$= -20x^9$$

✎ Pencil Problem #7 ✎

7. Multiply: $(2x^2)(-3x)(8x^4)$

Objective 6.R.8: Find the greatest common factor	
✔ *Solved Problem #8*	✎ *Pencil Problem #8* ✏

8a. Find the greatest common factor of the following list of monomials: $18x^3$ and $15x^2$

$18x^3 = 3x^2 \cdot 6x$
$15x^2 = 3x^2 \cdot 5$

The GCF is $3x^2$.

8a. Find the greatest common factor of the following list of monomials: $12x^2$ and $8x$

8b. Find the greatest common factor of the following list of monomials: x^4y, x^3y^2, and x^2y

$x^4y = x^2y \cdot x^2$
$x^3y^2 = x^2y \cdot xy$
$x^2y = x^2y$

The GCF is x^2y.

8b. Find the greatest common factor of the following list of monomials: $16x^5y^4$, $8x^6y^3$, and $20x^4y^5$

504

Objective 6.R.9: Use the product rule	

✔ **Solved Problem #9**

9a. Multiply using the product rule: $b^6 \cdot b^5$

$b^6 \cdot b^5 = b^{6+5}$
$\qquad = b^{11}$

✎ **Pencil Problem #9** ✎

9a. Multiply using the product rule: $3x^4 \cdot 2x^2$

9b. Multiply using the product rule: $\left(4x^3 y^4\right)\left(10x^2 y^6\right)$

$\left(4x^3 y^4\right)\left(10x^2 y^6\right) = 4 \cdot 10 \cdot x^3 \cdot x^2 \cdot y^4 \cdot y^6$
$\qquad\qquad = 40x^{3+2} y^{4+6}$
$\qquad\qquad = 40x^5 y^{10}$

9b. Multiply using the product rule: $\left(-2y^{10}\right)\left(-10y^2\right)$

Objective 6.R.10: Use the power rule

✔ Solved Problem #10

10. Simplify $(b^{-3})^{-4}$ using the power rule.

$(b^{-3})^{-4} = b^{(-3)(-4)}$

$\quad\quad\quad = b^{12}$

✎ Pencil Problem #10 ✎

10. Simplify $\left(7^{-4}\right)^{-5}$ using the power rule.

Objective 6.R.11: Use the quotient rule

✔ Solved Problem #11

11. Divide using the quotient rule: $\dfrac{27x^{14}y^8}{3x^3y^5}$

$$\dfrac{27x^{14}y^8}{3x^3y^5} = \dfrac{27}{3}x^{14-3}y^{8-5}$$

$$= 9x^{11}y^3$$

✎ Pencil Problem #11 ✎

11. Divide using the quotient rule: $\dfrac{50x^2y^7}{5xy^4}$

Objective 6.R.12: Simplify exponential expressions

✔ *Solved Problem*	✏ *Pencil Problem#8* ✏

8a. Simplify: $\left(-3x^{-6}y\right)\left(-2x^3y^4\right)^2$

$$(-3x^{-6}y)(-2x^3y^4)^2 = (-3x^{-6}y)(-2)^2(x^3)^2(y^4)^2$$
$$= -3 \cdot x^{-6} \cdot y \cdot 4 \cdot x^6 \cdot y^8$$
$$= -12 \cdot x^{-6+6} \cdot y^{1+8}$$
$$= -12x^0 y^9$$
$$= -12y^9$$

8a. Simplify: $\left(2a^5\right)\left(-3a^{-7}\right)$

8b. Simplify: $\left(\dfrac{10x^3y^5}{5x^6y^{-2}}\right)^2$

$$\left(\frac{10x^3y^5}{5x^6y^{-2}}\right)^2 = \left(2x^{3-6}y^{5+2}\right)^2$$
$$= \left(2x^{-3}y^7\right)^2$$
$$= 4x^{-6}y^{14}$$
$$= \frac{4y^{14}}{x^6}$$

8b. Simplify: $\left(\dfrac{-15a^4b^2}{5a^{10}b^{\,3}}\right)^3$

508

<u>Answers for Pencil Problems</u>:

1a. solution   **1b.** not a solution   **2a.** $\{23\}$ **2b.** $\{4\}$ **2c.** $\left\{-\dfrac{17}{12}\right\}$ **2d.** $\{11\}$ **2e.** $\{6\}$

2f. $\{-12\}$ **3a.** $4x = 28$ **3b.** $5x = 24 - x$ **4a.** $\{-1\}$ **4b.** $\{6\}$

5a.

5b.

6a. $17x$ **6b.** $8a$ **6c.** $14a + 14$ **7.** $-48x^7$ **8a.** $4x$ **8b.** $4x^4 y^3$ **9a.** $6x^6$ **9b.** $20y^{12}$ **10.** 7^{20}

11. $10xy^3$ **12a.** $-\dfrac{6}{a^2}$ **12b.** $-\dfrac{27b^{15}}{a^{18}}$

CHAPTER 7 – Algebra: Graphs, Functions, and Linear Systems

Guided Practice:

☐ Review each of the following **Solved Problems** and complete each **Pencil Problem**.

Objective 7.R.1: Determine whether an ordered pair is a solution of an equation

✔ **Solved Problem #1**	✎ **Pencil Problem #1** ✎
1a. Determine whether the ordered pair $(3, -2)$ is a solution of the equation $x - 3y = 9$.	**1a.** Determine whether the ordered pair $(0, 6)$ is a solution of the equation $y = 2x + 6$.

$x - 3y = 9$
$3 - 3(-2) = 9$
$3 + 6 = 9$
$9 = 9$, true

$(3, -2)$ is a solution.

1b. Determine whether the ordered pair $(-2, 3)$ is a solution of the equation $x - 3y = 9$.

$x - 3y = 9$
$-2 - 3(3) = 9$
$-2 - 9 = 9$
$-11 = 9$, false

$(-2, 3)$ is not a solution.

1b. Determine whether the ordered pair $(2, -2)$ is a solution of the equation $y = 2x + 6$.

510

✔ *Solved Problem #2*    *Pencil Problem #2* ✏

2. Find five solutions of $y = 3x + 2$.

Select integers for x, starting with -2 and ending with 2.

x	$y = 3x + 2$	(x, y)
-2	$y = 3(-2) + 2$ $= -6 + 2$ $= -4$	$(-2, -4)$
-1	$y = 3(-1) + 2$ $= -3 + 2$ $= -1$	$(-1, -1)$
0	$y = 3(0) + 2$ $= 0 + 2$ $= 2$	$(0, 2)$
1	$y = 3(1) + 2$ $= 3 + 2$ $= 5$	$(1, 5)$
2	$y = 3(2) + 2$ $= 6 + 2$ $= 8$	$(2, 8)$

2. Find five solutions of $y = -3x + 7$.

Select integers for x, starting with -2 and ending with 2.

Objective 7.R.3: Use a graph to identify intercepts

✔ **Solved Problem #3**

3a. Identify the *x*- and *y*- intercepts:

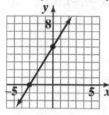

The graph crosses the *x*-axis at (–3,0).
Thus, the *x*-intercept is –3.

The graph crosses the *y*-axis at (0,5).
Thus, the *y*-intercept is 5.

3b. Identify the *x*- and *y*- intercepts:

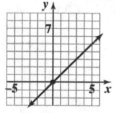

The graph crosses the *x*-axis at (0,0).
Thus, the *x*-intercept is 0.

The graph crosses the *y*-axis at (0,0).
Thus, the *y*-intercept is 0.

✎ **Pencil Problem #3** ✎

3a. Identify the *x*- and *y*- intercepts:

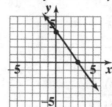

3b. Identify the *x*- and *y*- intercepts:

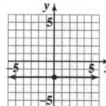

512

Objective 7.R.4: Graph equations in the rectangular coordinate system

| ✔ **Solved Problem #4** | ✏ **Pencil Problem #4** ✏ |

4a. Graph $y = 1 - x^2$.

x	$y = 1 - x^2$	(x, y)
-3	$y = 1 - (-3)^2 = -8$	$(-3, -8)$
-2	$y = 1 - (-2)^2 = -3$	$(-2, -3)$
-1	$y = 1 - (-1)^2 = 0$	$(-1, 0)$
0	$y = 1 - (0)^2 = 1$	$(0, 1)$
1	$y = 1 - (1)^2 = 0$	$(1, 0)$
2	$y = 1 - (2)^2 = -3$	$(2, -3)$
3	$y = 1 - (3)^2 = -8$	$(3, -8)$

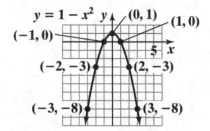

4a. Graph $y = x^2 - 4$. Let $x = -3, -2, -1, 0, 1, 2,$ and 3.

4b. Graph $y = |x + 1|$.

x	$y = |x + 1|$	(x, y)
-4	$y = |-4 + 1| = |-3| = 3$	$(-4, 3)$
-3	$y = |-3 + 1| = |-2| = 2$	$(-3, 2)$
-2	$y = |-2 + 1| = |-1| = 1$	$(-2, 1)$
-1	$y = |-1 + 1| = |0| = 0$	$(-1, 0)$
0	$y = |0 + 1| = |1| = 1$	$(0, 1)$
1	$y = |1 + 1| = |2| = 2$	$(1, 2)$
2	$y = |2 + 1| = |3| = 3$	$(2, 3)$

4b. Graph $y = 2|x|$. Let $x = -3, -2, -1, 0, 1, 2,$ and 3.

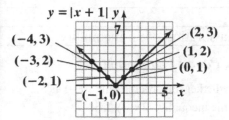

$y = |x + 1|$

$(-4, 3)$
$(-3, 2)$
$(-2, 1)$
$(-1, 0)$
$(0, 1)$
$(1, 2)$
$(2, 3)$

Objective 7.R.5: Solve a formula for a variable

✔ **Solved Problem #7**

5a. Solve the formula $2l + 2w = P$ for w.

$2l + 2w = P$

$2w = P - 2l$

$\dfrac{2w}{2} = \dfrac{P - 2l}{2}$

$w = \dfrac{P - 2l}{2}$

✎ **Pencil Problem #5** ✎

5a. Solve the formula $T = D + pm$ for p.

5b. Solve the formula $P = C + MC$ for C.

$P = C + MC$

$P = C(1 + M)$

$\dfrac{P}{1 + M} = \dfrac{C(1 + M)}{1 + M}$

$\dfrac{P}{1 + M} = C$

$C = \dfrac{P}{1 + M}$

5b. Solve the formula $IR + Ir = E$ for I.

514

✔ Solved Problem #6

6a. Determine whether the ordered pair $(0,0)$ is a solution of the inequality $5x + 4y \leq 20$.

Substitute 0 for x and 0 for y into the inequality.
$$5x + 4y \leq 20$$
$$5(0) + 4(0) \leq 20$$
$$0 \leq 20, \text{ true}$$

$(0,0)$ is a solution.

✎ Pencil Problem #6 ✎

6a. Determine whether the ordered pair $(0,0)$ is a solution of the inequality $2x + y \geq 5$.

6b. Determine whether the ordered pair $(6,2)$ is a solution of the inequality $5x + 4y \leq 20$.

Substitute 6 for x and 2 for y into the inequality.
$$5x + 4y \leq 20$$
$$5(6) + 4(2) \leq 20$$
$$30 + 8 \leq 20$$
$$38 \leq 20, \text{ false}$$

$(6,2)$ is not a solution.

6b. Determine whether the ordered pair $(-2,-4)$ is a solution of the inequality $y \geq -2x + 4$.

Objective 7.R.7: Understand properties used to solve linear inequalities

✔ *Solved Problem #7*	 *Pencil Problem #7*
7. *True* or *False*: When we add (or subtract) a negative number to (or from) both sides of an inequality, the direction of the inequality symbol is reversed.	7. *True* or *False*: When we multiply or divide both sides of an inequality by a negative number, the direction of the inequality symbol is reversed.

False; This rule applies to multiplication and division.

516

1a. $(0,6)$ is a solution

1b. $(2,-2)$ is not a solution

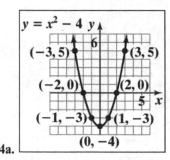

x	$y = -3x + 7$	(x, y)
-2	$y = -3(-2) + 7 = 13$	$(-2, 13)$
-1	$y = -3(-1) + 7 = 10$	$(-1, 10)$
0	$y = -3(0) + 7 = 7$	$(0, 7)$
1	$y = -3(1) + 7 = 4$	$(1, 4)$
2	$y = -3(2) + 7 = 1$	$(2, 1)$

2.

3a. x-intercept is 3 and y-intercept is 4

3b. no x-intercept and y-intercept is -2

4a.

4b.

5a. $p = \dfrac{T - D}{m}$

5b. $I = \dfrac{E}{R + r}$

6a. not a solution

6b. not a solution

7. True   *(2.7 #45,47)*

CHAPTER 8 – Personal Finance

Guided Practice:

☐ Review each of the following *Solved Problems* and complete each *Pencil Problem*.

Objective 8.R.1: Use the percent formula	
✔ **Solved Problem #1**	✎ **Pencil Problem #1** ✎
1a. What is 9% of 50?	**1a.** What is 3% of 200?
Use the formula $A = PB$: A is P percent of B.	
What is 9% of 50? $$A = 0.09 \cdot 50$$ $$A = 4.5$$	
4.5 is 9% of 50.	
1b. 9 is 60% of what?	**1b.** 24% of what number is 40.8?
Use the formula $A = PB$: A is P percent of B.	
9 is 60% of what? $$9 = 0.60 \cdot B$$ $$\frac{9}{0.60} = \frac{0.60B}{0.60}$$ $$15 = B$$	
9 is 60% of 15.	
1c. 18 is what percent of 50?	**1c.** 3 is what percent of 15?
Use the formula $A = PB$: A is P percent of B.	
18 is what percent of 50? $$18 = P \cdot 50$$ $$18 = P \cdot 50$$ $$\frac{18}{50} = \frac{50P}{50}$$ $$0.36 = P$$ To change 0.36 to a percent, move the decimal point two places to the right and add a percent sign. $$0.36 = 36\%$$	
18 is 36% of 50.	

518

Answers for Pencil Problems

1a. 6   **1b.** 170   **1c.** 20%

CHAPTER 9 – Measurement

Guided Practice:

☐ Review each of the following *Solved Problems* and complete each *Pencil Problem*.

Objective 9.R.1: Convert between mixed numbers and improper fractions	
✔ **Solved Problem #1**	✎ **Pencil Problem #1**

1a. Convert $2\frac{5}{8}$ to an improper fraction.

$$2\frac{5}{8} = \frac{2 \cdot 8 + 5}{8}$$
$$= \frac{16 + 5}{8}$$
$$= \frac{21}{8}$$

1a. Convert $7\frac{3}{5}$ to an improper fraction.

1b. Convert $\frac{5}{3}$ to a mixed number.

5 divided by 3 is 1 with a remainder of 2,
so $\frac{5}{3} = 1\frac{2}{3}$.

1b. Convert $\frac{76}{9}$ to a mixed number.

Objective 9.R.2: Convert from decimal to scientific notation	
✔ **Solved Problem #2**	✎ **Pencil Problem #2**

2a. Write in scientific notation: $5,210,000,000$

The decimal point must be moved 9 places to the left to get a number whose absolute value is between 1 and 10. Thus the exponent on 10 is 9.

$$5,210,000,000 = 5.21 \times 10^9$$

2a. Write in scientific notation: $32,000$

2b. Write in scientific notation: -0.00000006893

The decimal point must be moved 8 places to the right to get a number whose absolute value is between 1 and 10. Thus the exponent on 10 is -8.

$$-0.00000006893 = -6.893 \times 10^{-8}$$

2b. Write in scientific notation: 0.0027

Objective 9.R.3: Convert from scientific to decimal notation	
✔ *Solved Problem #3*	✎ *Pencil Problem #3* ✎

3a. Write in decimal notation: -2.6×10^9

Move the decimal point 9 places to the right.

$-2.6 \times 10^9 = -2,600,000,000$

3a. Write in decimal notation: -7.16×10^6

3b. Write in decimal notation: 3.017×10^{-6}

Move the decimal point 6 places to the left.

$3.017 \times 10^{-6} = 0.000003017$

3b. Write in decimal notation: -4.15×10^{-3}

Answers for Pencil Problems:

1a. $\dfrac{38}{5}$ **1b.** $8\dfrac{4}{9}$ **2a.** 3.2×10^4 **2b.** 2.7×10^{-3} **3a.** $-7,160,000$ **3b.** -0.00415

CHAPTER 10 – Geometry

Guided Practice:

☐ Review each of the following *Solved Problems* and complete each *Pencil Problem*.

Objective 10.R.1: Solve problems involving proportions

✔ **Solved Problem #1**	✎ **Pencil Problem #1** ✎
1a. The property tax on a house with an assessed value of $250,000 is $3500. Determine the property tax on a house with an assessed value of $420,000, assuming the same tax rate.	**1a.** The tax on a property with an assessed value of $65,000 is $720. Find the tax on a property with an assessed value of $162,500.

Let $x =$ the property tax on the $420,000 house.

$$\frac{\text{Tax on \$250,000 house}}{\text{Assessed value (\$250,000)}} = \frac{\text{Tax on \$420,000 house}}{\text{Assessed value (\$420,000)}}$$

$$\frac{\$3500}{\$250,000} = \frac{\$x}{\$420,000}$$

$$\frac{3500}{250,000} = \frac{x}{420,000}$$

$$250,000x = (3500)(420,000)$$

$$250,000x = 1,470,000,000$$

$$\frac{250,000x}{250,000} = \frac{1,470,000,000}{250,000}$$

$$x = 5880$$

The property tax is $5880.

| **1b.** Wildlife biologists catch, tag, and then release 120 deer back into a wildlife refuge. Two weeks later they observe a sample of 150 deer, 25 of which are tagged. Assuming the ratio of tagged deer in the sample holds for all deer in the refuge, approximately how many deer are in the refuge? | **1b.** St. Paul Island in Alaska has 12 fur seal rookeries (breeding places). In 1961, to estimate the fur seal pup population in the Gorbath rookery, 4963 fur seal pups were tagged in early August. In late August, a sample of 900 pups was observed and 218 of these were found to have been previously tagged. Estimate the total number of fur seal pups in this rookery. |

Let $x =$ the total number of deer in the refuge.

$$\frac{120}{x} = \frac{25}{150}$$

$$25x = (120)(150)$$

$$25x = 18,000$$

$$\frac{25x}{25} = \frac{18,000}{25}$$

$$x = 720$$

There are about 720 deer in the refuge.

Objective 10.R.2: Solve quadratic equations using the square root property

✔ *Solved Problem #2*	✏ *Pencil Problem #2* ✏
2a. Solve by the square root property: $x^2 = 36$	**2a.** Solve by the square root property: $y^2 = 81$

$x^2 = 36$
$x = \sqrt{36}$ or $x = -\sqrt{36}$
$x = 6 \qquad x = -6$

The solution set is $\{\pm 6\}$.

2b. Solve by the square root property: $5x^2 = 15$

2b. Solve by the square root property: $4y^2 = 49$

$5x^2 = 15$

$\dfrac{5x^2}{5} = \dfrac{15}{5}$

$x^2 = 3$

$x = \sqrt{3}$ or $x = -\sqrt{3}$

The solution set is $\left\{\pm\sqrt{3}\right\}$.

2c. Solve by the square root property: $2x^2 - 7 = 0$

2c. Solve by the square root property: $2x^2 + 1 = 51$

$2x^2 - 7 = 0$

$\quad 2x^2 = 7$

$\quad\quad x^2 = \dfrac{7}{2}$

$x = \sqrt{\dfrac{7}{2}}$ or $x = -\sqrt{\dfrac{7}{2}}$

$x = \pm\sqrt{\dfrac{7}{2}}$

$\quad = \pm\dfrac{\sqrt{7}}{\sqrt{2}} \cdot \dfrac{\sqrt{2}}{\sqrt{2}}$

$\quad = \pm\dfrac{\sqrt{14}}{2}$

The solution set is $\left\{\pm\dfrac{\sqrt{14}}{2}\right\}$.

524

2d. Solve by the square root property: $(x-3)^2 = 25$

$$(x-3)^2 = 25$$
$$x - 3 = \sqrt{25} \quad \text{or} \quad x - 3 = -\sqrt{25}$$
$$x - 3 = 5 \quad \text{or} \quad x - 3 = -5$$
$$x = 8 \qquad \qquad x = -2$$

The solution set is $\{-2, 8\}$.

2d. Solve by the square root property: $(x+5)^2 = 121$

2e. Solve by the square root property: $(x-2)^2 = 7$

$$(x-2)^2 = 7$$
$$x - 2 = \sqrt{7} \quad \text{or} \quad x - 2 = -\sqrt{7}$$
$$x = 2 + \sqrt{7} \qquad x = 2 - \sqrt{7}$$

The solution set is $\{2 \pm \sqrt{7}\}$.

2e. Solve by the square root property: $(x-5)^2 = 3$

Answers for Pencil Problems

1a. $1800 **1b.** about 20,489 fur seal pups **2a.** $\{\pm 9\}$ **2b.** $\left\{\pm \dfrac{7}{2}\right\}$ **2c.** $\{\pm 5\}$ **2d.** $\{-16, 6\}$

2e. $\left\{5 \pm \sqrt{3}\right\}$

CHAPTER 11 – Counting Methods and Probability Theory

Guided Practice:

☐ Review each of the following *Solved Problems* and complete each *Pencil Problem*.

Objective 11.R.1: Reduce or simplify fractions

✔ *Solved Problem #1*	✎ *Pencil Problem #1* ✎
1a. Reduce $\dfrac{10}{15}$ to its lowest terms.	**1a.** Reduce $\dfrac{35}{50}$ to its lowest terms.

$$\frac{10}{15} = \frac{2 \cdot \cancel{5}}{3 \cdot \cancel{5}} = \frac{2}{3}$$

1b. Reduce $\dfrac{13}{15}$ to its lowest terms.

13 and 15 share no common factors (other than 1).

Therefore, $\dfrac{13}{15}$ is already reduced to its lowest terms.

1b. Reduce $\dfrac{120}{86}$ to its lowest terms.

Objective 11.R.2: Express rational numbers as decimals

✔ *Solved Problem #2*	✎ *Pencil Problem #2* ✎
2a. Express the rational number as a decimal: $\dfrac{3}{8}$	**2a.** Express the rational number as a decimal: $\dfrac{7}{8}$

$$
\begin{array}{r}
0.375 \\
8\overline{)3.000} \\
\underline{24} \\
60 \\
\underline{56} \\
40 \\
\underline{40} \\
0
\end{array}
$$

$\dfrac{3}{8} = 0.375$

2b. Express the rational number as a decimal: $\dfrac{5}{11}$

2b. Express the rational number as a decimal: $\dfrac{9}{11}$

$$
\begin{array}{r}
0.454... \\
11\overline{)5.000...} \\
\underline{44} \\
60 \\
\underline{55} \\
50 \\
\underline{44} \\
60
\end{array}
$$

$\dfrac{5}{11} = 0.\overline{45}$

Objective 11.R.3: Multiply Fractions	
✔ *Solved Problem #3*	✎ *Pencil Problem #3* ✎

3a. Multiply $\dfrac{4}{11} \cdot \dfrac{2}{3}$.

If possible, reduce the product to its lowest terms.

$$\dfrac{4}{11} \cdot \dfrac{2}{3} = \dfrac{4 \cdot 2}{11 \cdot 3}$$
$$= \dfrac{8}{33}$$

3a. Multiply $\dfrac{3}{8} \cdot \dfrac{7}{11}$.

If possible, reduce the product to its lowest terms.

3b. Multiply $\left(3\dfrac{2}{5}\right)\left(1\dfrac{1}{2}\right)$.

If possible, reduce the product to its lowest terms.

$$\left(3\dfrac{2}{5}\right)\left(1\dfrac{1}{2}\right) = \dfrac{17}{5} \cdot \dfrac{3}{2}$$
$$= \dfrac{51}{10}$$
$$= 5\dfrac{1}{10}$$

3b. Multiply $\left(3\dfrac{3}{4}\right)\left(1\dfrac{3}{5}\right)$.

If possible, reduce the product to its lowest terms.

Objective 11.R.4: Add and subtract fractions with identical denominators	
✔ *Solved Problem #4*	✎ *Pencil Problem #4* ✏

4a. Perform the indicated operation: $\dfrac{2}{11} + \dfrac{3}{11}$

$$\dfrac{2}{11} + \dfrac{3}{11} = \dfrac{2+3}{11}$$
$$= \dfrac{5}{11}$$

4a. Perform the indicated operation: $\dfrac{7}{12} + \dfrac{1}{12}$

4b. Perform the indicated operation: $\dfrac{5}{6} - \dfrac{1}{6}$

$$\dfrac{5}{6} - \dfrac{1}{6} = \dfrac{4}{6}$$
$$= \dfrac{2}{3}$$

4b. Perform the indicated operation: $\dfrac{11}{18} - \dfrac{4}{18}$

530

Objective 11.R.5: Add and subtract fractions with unlike denominators	
✔ *Solved Problem #5*	✏ *Pencil Problem #5* ✏

5a. Perform the indicated operation: $\dfrac{1}{2}+\dfrac{3}{5}$

$$\dfrac{1}{2}+\dfrac{3}{5}=\dfrac{1\cdot5}{2\cdot5}+\dfrac{3\cdot2}{5\cdot2}$$
$$=\dfrac{5}{10}+\dfrac{6}{10}$$
$$=\dfrac{5+6}{10}$$
$$=\dfrac{11}{10}$$

5a. Perform the indicated operation: $\dfrac{3}{8}+\dfrac{5}{12}$

5b. Perform the indicated operation: $3\dfrac{1}{6}-1\dfrac{11}{12}$

$$3\dfrac{1}{6}-1\dfrac{11}{12}=\dfrac{19}{6}-\dfrac{23}{12}$$
$$=\dfrac{19\cdot2}{6\cdot2}-\dfrac{23}{12}$$
$$=\dfrac{38}{12}-\dfrac{23}{12}$$
$$=\dfrac{15}{12}$$
$$=\dfrac{5}{4}$$
$$=1\dfrac{1}{4}$$

5b. Perform the indicated operation: $3\dfrac{3}{4}-2\dfrac{1}{3}$

Objective 11.R.6: Divide fractions

✔ *Solved Problem #6*	✏ *Pencil Problem #6* ✏

6a. Divide $\dfrac{5}{4} \div \dfrac{3}{8}$.

$$\dfrac{5}{4} \div \dfrac{3}{8} = \dfrac{5}{4} \cdot \dfrac{8}{3}$$

$$= \dfrac{5}{\cancel{4}} \cdot \dfrac{\cancel{4} \cdot 2}{3}$$

$$= \dfrac{10}{3}$$

$$= 3\dfrac{1}{3}$$

6a. Divide $\dfrac{7}{6} \div \dfrac{5}{3}$.

6b. Divide $3\dfrac{3}{8} \div 2\dfrac{1}{4}$.

$$3\dfrac{3}{8} \div 2\dfrac{1}{4} = \dfrac{27}{8} \div \dfrac{9}{4}$$

$$= \dfrac{27}{8} \cdot \dfrac{4}{9}$$

$$= \dfrac{\cancel{9} \cdot 3}{\cancel{4} \cdot 2} \cdot \dfrac{\cancel{4}}{\cancel{9}}$$

$$= \dfrac{3}{2}$$

$$= 1\dfrac{1}{2}$$

6b. Divide $6\dfrac{3}{5} \div 1\dfrac{1}{10}$.

532

1a. $\frac{7}{10}$ **1b.** $\frac{60}{43}$ **2a.** 0.875   **2b.** $0.\overline{81}$ **3a.** $\frac{21}{88}$ **3b.** 6   **4a.** $\frac{2}{3}$ **4b.** $\frac{7}{18}$ **5a.** $\frac{19}{24}$ **5b.** $\frac{17}{12}$ or $1\frac{5}{12}$

6a. $\frac{7}{10}$ **6b.** 6

CHAPTER 12 – Statistics

Guided Practice:

☐ Review each of the following *Solved Problems* and complete each *Pencil Problem*.

Objective 12.R.1: Plot ordered pairs in the rectangular coordinate system

✔ *Solved Problem #1*

1. Plot the points:
   $A(-2, 4)$, $B(4, -2)$, $C(-3, 0)$, and $D(0, -3)$.

From the origin, point A is left 2 units and up 4 units.

From the origin, point B is right 4 units and down 2 units.

From the origin, point C is left 3 units.

From the origin, point D is down 3 units.

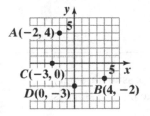

✏ *Pencil Problem #1* ✏

1. Plot the points:
   $A(3, 5)$, $B(-5, 1)$, $C(-3, -1)$.

Objective 12.R.2: Compute a line's slope

✔ *Solved Problem #2*

2a. Find the slope of the line passing through the pair of points: $(4,-2)$ and $(-1,5)$.

Let $(x_1, y_1) = (4,-2)$ and $(x_2, y_2) = (-1,5)$.

$$m = \frac{\text{Change in } y}{\text{Change in } x} = \frac{y_2 - y_1}{x_2 - x_1} = \frac{5-(-2)}{-1-4} = \frac{7}{-5} = -\frac{7}{5}$$

The slope is $-\dfrac{7}{5}$.

Since the slope is negative, the line falls from left to right.

2b. Find the slope of the line passing through $(6,5)$ and $(2,5)$ or state that the slope is undefined. Indicate if the line is horizontal or vertical.

Let $(x_1, y_1) = (6,5)$ and $(x_2, y_2) = (2,5)$.

$$m = \frac{\text{Change in } y}{\text{Change in } x} = \frac{y_2 - y_1}{x_2 - x_1}$$
$$= \frac{5-5}{2-6}$$
$$= \frac{0}{-4}$$
$$= 0$$

Since the slope is 0, the line is horizontal.

✎ *Pencil Problem #2* ✎

2a. Find the slope of the line passing through the pair of points: $(4,7)$ and $(8,10)$.

2b. Find the slope of the line passing through $(4,-2)$ and $(3,-2)$ or state that the slope is undefined.
Indicate if the line is horizontal or vertical.

2c. Find the slope of the line passing through $(1,6)$ and $(1,4)$ or state that the slope is undefined. Indicate if the line is horizontal or vertical.

Let $(x_1, y_1) = (1,6)$ and $(x_2, y_2) = (1,4)$.

$$m = \frac{\text{Change in } y}{\text{Change in } x} = \frac{y_2 - y_1}{x_2 - x_1}$$

$$= \frac{4-6}{1-1}$$

$$= \frac{-2}{0}$$

Because division by 0 is undefined, the slope is undefined.

Since the slope is undefined, the line is vertical.

2c. Find the slope of the line passing through $(5,3)$ and $(5,-2)$ or state that the slope is undefined. Indicate if the line is horizontal or vertical.

536

✔ *Solved Problem #3*

✎ *Pencil Problem #3*✎

3a. Find the slope and the y-intercept of the line:
$$y = \frac{2}{3}x + 4$$

The slope is the x-coefficient, which is $m = \frac{2}{3}$.

The y-intercept is the constant term, which is 4.

3a. Find the slope and the y-intercept of the line:
$$y = -\frac{1}{2}x + 5$$

3b. Find the slope and the y-intercept of the line:
$$7x + y = 6$$

First, solve the equation for y.
$$7x + y = 6 \;\rightarrow\; y = -7x + 6$$

The slope is the x-coefficient, which is $m = -7$.

The y-intercept is the constant term, which is 6.

3b. Find the slope and the y-intercept of the line:
$$3x + 2y = 3$$

Objective 12.R.4: Graph lines in slope-intercept form	
✔ **Solved Problem #4**	✏ *Pencil Problem #4* ✏

4a. Graph: $y = 3x - 2$

The *y*-intercept is –2, so plot the point $(0, -2)$.

The slope is $m = 3$ or $m = \dfrac{3}{1}$.

Find another point by going up 3 units and to the right 1 unit.

Use a straightedge to draw a line through the two points.

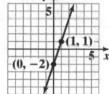

$y = 3x - 2$

4a. Graph: $y = 2x + 4$

4b. Graph: $y = \dfrac{3}{5}x + 1$

The *y*-intercept is 1, so plot the point $(0, 1)$.

The slope is $m = \dfrac{3}{5}$.

Find another point by going up 3 units and to the right 5 units.

Use a straightedge to draw a line through the two points.

4b. Graph: $y = -\dfrac{3}{4}x + 2$

538

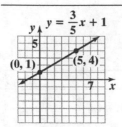

$$y = \frac{3}{5}x + 1$$

(0, 1)   (5, 4)

Objective 12.R.5: Use the rectangular coordinate system to visualize relationships between variables

✔ Solved Problem #5

5. When a physician injects a drug into a patient's muscle, the concentration of the drug in the body, measured in milligrams per 100 milliliters, depends on the time elapsed after the injection, measured in hours. The following figure shows the graph of drug concentration over time, where x represents hours after the injection and y represents the drug concentration at time x.

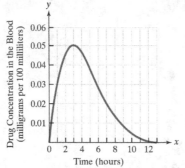

Use the figure to answer the following questions.

5a. During which period of time is the drug concentration increasing?

The drug concentration is increasing from 0 to 3 hours.

5b. During which period of time is the drug concentration decreasing?

The drug concentration is decreasing from 3 to 13 hours.

5c. What is the drug's maximum concentration and when does this occur?

The drug's maximum concentration is 0.05 milligram per 100 milliliters, which occurs after 3 hours.

✐ Pencil Problem #5 ✐

5. Select the graph that best illustrates the story:

An airplane flew from Miami to San Francisco.

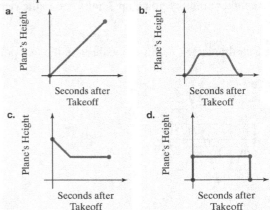

Copyright © 2015 Pearson Education Inc.

5d. What happens by the end of 13 hours?

None of the drug is left in the body.

Objective #12.R.6: Find a number's absolute value.

✔ *Solved Problem #1*	✎ *Pencil Problem #1*
6. Find the absolute value: $\lvert -6 \rvert$	**6.** Find the absolute value: $\lvert -7.6 \rvert$
-6 is 6 units from 0. Thus, $\lvert -6 \rvert = 6$	

540

Answers for Pencil Problems:

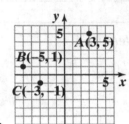

1.

2a. $m = \dfrac{3}{4}$

2b. $m = 0$; the line is horizontal

2c. the slope is undefined; the line is vertical

3a. $m = -\dfrac{1}{2}$ and the y-intercept is 5

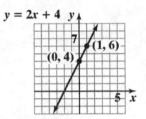

3b. $m = -\dfrac{3}{2}$ and the y-intercept is $\dfrac{3}{2}$

4a.

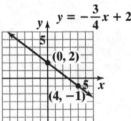

4b.

5. graph b

6. 7.6